William Cawthorne Unwin

On the Development and transmission of Power from central

Stations

William Cawthorne Unwin

On the Development and transmission of Power from central Stations

ISBN/EAN: 9783337024673

Printed in Europe, USA, Canada, Australia, Japan

Cover: Foto ©berggeist007 / pixelio.de

More available books at **www.hansebooks.com**

ON THE

DEVELOPMENT AND TRANSMISSION
OF
POWER

FROM CENTRAL STATIONS

BEING

The Howard Lectures

DELIVERED AT THE SOCIETY OF ARTS IN 1893

BY

WILLIAM CAWTHORNE UNWIN, F.R.S.

B.SC., M. INST. CIVIL ENGINEERS
M. INST. MECH. ENGINEERS, MEM. AM. PHILOSOPHICAL SOCIETY AND HON. MEM. FRANKLIN INST.
PROFESSOR OF ENGINEERING AT THE GUILDS CENTRAL TECHNICAL COLLEGE
FORMERLY PROFESSOR OF HYDRAULIC AND MECHANICAL ENGINEERING
AT THE ROYAL INDIAN ENGINEERING COLLEGE

LONDON
LONGMANS, GREEN, AND CO.
AND NEW YORK: 15 EAST 16th STREET
1894

PREFACE

THE present Treatise is based on the Course of Lectures which the Council of the Society of Arts requested the Author to deliver in January and February 1893. In republishing them under less stringent limitations of time and space many gaps have been filled up and some questions have been discussed more fully. The importance of the problem of distributing power to many consumers can hardly be overrated. In dealing with it the question of cost cannot be put on one side. The financial conditions are governing conditions, and must be considered together with the mechanical conditions. An attempt has been made in the present treatise to treat the subject as a whole. Hence the causes of waste in generating power have been discussed as well as the losses in distribution. The subject is so wide and touches so many departments of engineering that it is too much to hope that all the questions involved have been examined with sufficiently adequate knowledge.

But much care has been taken to indicate what is essential in the consideration of schemes of power distribution, whatever the source from which the power is obtained and whatever the method of transmission adopted. For the rest, practical experience will gradually determine much that is at present doubtful.

MAY, 1894.

185003

CONTENTS

CHAPTER   PAGE

I. THE CONDITIONS IN WHICH A SYSTEM OF DISTRIBUTION OF ENERGY IS REQUIRED, AND GENERAL CONSIDERATIONS ON THE SOURCES OF ENERGY 1

II. POWER GENERATED BY STEAM ENGINES. CONDITIONS OF ECONOMY AND WASTE 19

III. THE COST OF STEAM POWER 58

IV. THE STORAGE OF ENERGY. 67

V. WATER POWER 80

VI. HYDRAULIC MOTORS 95

VII. TELODYNAMIC TRANSMISSION 108

VIII. HYDRAULIC TRANSMISSION 129

IX. TRANSMISSION OF POWER BY COMPRESSED AIR . . . 163

X. CALCULATION OF A COMPRESSED AIR TRANSMISSION WHEN THE SUBSIDIARY LOSSES OF ENERGY ARE TAKEN INTO ACCOUNT 211

XI. DISTRIBUTION OF POWER BY STEAM 245

XII. DISTRIBUTION OF GAS FOR POWER PURPOSES . . . 255

XIII. ELECTRICAL TRANSMISSION OF POWER 263

XIV. EXAMPLES OF POWER TRANSMISSION BY ELECTRICAL METHODS 286

XV. THE UTILISATION OF NIAGARA FALLS 297

INDEX 309

DISTRIBUTION OF POWER

CHAPTER I

THE CONDITIONS IN WHICH A SYSTEM OF DISTRIBUTION
OF ENERGY IS REQUIRED, AND GENERAL CONSIDERA-
TIONS ON THE SOURCES OF ENERGY

THE late Mr. Thomas Howard left a bequest to the Society of
Arts, to provide for the delivery, periodically, of courses of
lectures relating to the production and use of motive power.

In carrying out the duty imposed by this trust, the Society
did the author the honour of inviting him to give some lectures
on the development of power at Central Stations, and its distribution, either as motive power for driving factories and
workshops, or as energy applied to other purposes. Energy is
in these lectures to be considered as a commodity, which can be
manufactured in a convenient form, and distributed and sold.
The special problems to be dealt with are the conditions which
favour the production of a convenient form of energy, on a large
scale and in the most economical way; the means of conveying
it to a distance and distributing it to consumers; the arrangements for measuring the quantity delivered; and lastly, the
relative advantages and disadvantages of a system in which
energy is obtained on a large scale and distributed to many
consumers, compared with a system in which each consumer
produces the power he requires in his own locality and under
his own supervision and responsibility.

As various sources of energy are available, and as there are
several methods of distributing energy or means of readily
obtaining energy, the inquiry is a wide one. It may be limited

B

at once to those cases in which the final form of energy required is mechanical energy. Pressure water is a convenient source of mechanical energy, so that a water-supply system with its pumping station and distributing mains is from one point of view a central station from which energy is distributed. But water-pressure systems need only be considered in these lectures so far as they are actually used for distributing motive power. An electric lighting station is a central station distributing energy to many consumers. But it will only come within the scope of these lectures to consider electrical systems, in which the whole or part of the electricity distributed is used for motive power purposes. The addition of motive power supply to the primary object of water supply or light supply may be of importance, and it is only so far as a combination of purposes in such systems facilitates or cheapens the supply of motive power that they will be considered.

That the author has more knowledge of some of the older and more mechanical methods of power distribution than of the newer methods of electrical distribution, is no doubt a disadvantage. But the distribution of electricity both for lighting and for power purposes has already been fully discussed, both in special treatises and especially in the lectures of Prof. George Forbes and Mr. Gisbert Kapp. On the other hand, there may be some partly compensating advantage in approaching the subject for once with the bias of an engineer rather than that of an electrician. Granting that electrical distribution will play an important part before long in the development of systems of power distribution, there is a popular tendency at the moment to regard too exclusively electrical methods, and to overlook other means of power distribution which have been usefully applied in the past, and will, in suitable conditions, be still employed in the future.

Unscientific people who see the electric lamp are perhaps a little apt to attribute to electricity too exclusive a share in the production of the light. They forget the coal, the boilers, and the steam engines, which are as necessary as the dynamo in obtaining the result. Perhaps, with less unscientific people, the striking success achieved in transmitting energy electrically to great distances, and retransforming it when necessary into mechanical energy, has obscured the fact that there are other methods of power transmission more convenient and less costly in particular cases.

Two points should be clearly kept in mind. First, that, as to the production of energy in an available form, we are just where we were before modern electrical discoveries were made. The most the electrician can do is to provide a new mechanism of distribution. The case would be different if, by means of primary batteries employing new materials for chemical action, or by thermo-electric batteries in which heat is directly transformed into electricity, the methods of producing available energy were changed. Such things are theoretically possible. Practically they are not yet available. At present the energy to be distributed must be developed by a steam-engine or water-wheel, and the dynamo, cable and electric motor merely replace the shafting, gearing and belting, or other mechanism of transmission, performing the same functions, more cheaply and effectively in certain cases it is true.

The second point is that every method of transmission will be found to have some characteristic advantages fitting it specially for particular cases. It may be conceded to the electrician that the special advantages of electrical transmission are strikingly apparent, where power must be conveyed to very great distances. But such cases are likely to be comparatively rare. The remarkable mechanical and scientific success of the Frankfort-Lauffen experiment, in which 300 h.p. was conveyed 108 miles with a loss in transmission of only 25 per cent., has a little misled merely popular observers. The fact must be borne in mind that the cost of the power, when it reached Frankfort, was five times as great as that of an equal amount of power produced directly in Frankfort by a steam-engine. It can only be, where some exceptionally cheap source of power at a distance can be put in competition with dearer power near at hand, by means of electrical transmission, that long-distance transmission is likely to be applied. For transmissions to moderate distances there is a choice of several means of transmission, and electrical distribution has not in such cases and up to the present time established any universal superiority.

The cost of production of mechanical energy from natural sources has to be considered in these lectures for several reasons.

If systems of producing energy in a wholesale way at central stations and for its distribution in parcels to different consumers should be adopted, it will chiefly be because energy can be so obtained from sources so much cheaper, or at so much less cost of production, that it can be distributed and sold at less cost than that at which it can be produced by individual consumers. It is necessary, therefore, to consider in what ways central station working renders available new sources of energy or is favourable to economy—what, in fact, is the relative cost of energy obtained in central stations on a large scale and in isolated workshops on a smaller scale. The methods and cost of distribution form a second branch of the inquiry.

General Statement of the Advantages of Central Stations for Producing Energy.—Some preliminary general considerations may be stated in favour of systems of producing energy at central stations and distributing it.

1. In generating power by steam-engines and boilers there is obviously some advantage in cost of machinery, in economy of fuel, and in cost of superintendence due to concentration of the engines and boilers in a single station. The central station can be placed where coal can be delivered cheaply, and arrangements can be adopted to secure economy which would be too costly or too complicated in small plants.

2. In the case of water power, it often happens that it is only possible to deal with a natural waterfall either by a combination of consumers, or by an association acting in the interest of many consumers, for the construction of the costly permanent works required.

3. The locality where the power can be generated may be fixed by one set of conditions, that where the power must be used by another set of conditions. Often it is a question of adopting a cheaper source of power at a distance or a dearer source near at hand. Thus, in mining and tunnelling operations, cheap steam power at the surface may be distributed to replace much more expensive hand labour in the workings. That is essentially a case of power distribution from a central station. Mr. Thwaite has recently proposed to erect large gas motor stations at collieries and to transmit the power electrically to the nearest industrial centres, where the power could be used in manufacturing operations.[1] Whether such a scheme is

[1] *The Engineer*, December 2, 1892.

advantageous or not is mainly a question of finance. The question to be solved is, whether there is so much economy in producing energy at the colliery that it can be delivered in the form required at a distant point, at a cost less than that at which it could be produced on the spot where it is used; for instance, from coal brought from the colliery by railway. In utilising water power, a similar problem often arises. The water power can only be utilised at a point distant from the manufactory where power is required. Then it has to be considered whether the transmitted water power costs more or less than power obtained at the manufactory from coal.

4. Another reason for central-station working, in large towns, is the great inconvenience of generating power in the small quantity and at the irregular times at which it is wanted for many purposes. For good or for ill, population gathers into huge communities, in which there is a complex development of social and industrial life. In such communities there is a constantly increasing need of mechanical power. In addition to manufacturing operations, demands for power arise for transit, for handling goods, for passenger lifts, for water supply, and for sanitation. At first these are met by the erection of scattered motors. But this sporadic production of power in small quantities is quite certainly in many instances extravagantly costly and inconvenient. There is a great probability that power distributed from a central station in towns would be so convenient as to be preferable to power produced locally, even at a somewhat greater cost.

Just as it has become necessary to supersede private systems of water supply by a municipal supply, just as it has proved convenient to distribute coal gas and necessary to establish a general system of sewerage, so it will probably be found convenient and even necessary to provide, in towns of a certain importance, some means of obtaining mechanical power in any desired quantity, and at a price proportional to the amount of power used. It is socialism in the field of mechanical engineering.

For the single purpose of working lifts and hoisting machinery, it has already proved remunerative to extend through the streets of London a system of hydraulic mains nearly sixty miles in length. The extremely rapid extension of the system is worthy of note. In July 1884, there were only 96 consumers taking power from the London Hydraulic Company's mains. In 1888, there were 720 consumers renting power. In 1892, there were 1,676 consumers renting power, and the use of the system is now extending more rapidly than at any previous period. The quantity of water distributed has increased from 317,000 gallons per week in 1884 to 6,600,000 gallons per week in 1892. In no instance has the use of hydraulic power when once adopted been abandoned in favour of any other system of working hoisting machinery.

There can be no doubt that when (a) power can be obtained with little trouble, in a form involving no great amount of superintendence in working, and (b) the cost is proportional to the amount of power used, then a demand for the power is readily created.

Perhaps a more striking instance of the growth of a demand for power is furnished by the town of Geneva. No casual observer would have selected Geneva, with its population of 50.000, as a likely centre for a great system of power distribution. Yet the works at Geneva, which will be more fully described later, are perhaps the most important example of power distribution hitherto carried out. In 1871, Colonel Turrettini obtained permission from the Municipal Council to place a water-pressure engine on the existing low-pressure town mains, for driving the factory of the Genevan Society for the Manufacture of Scientific Instruments. In case of this method of obtaining power proving satisfactory, he obtained the right to instal similar motors in other parts of the town. The plan proved so convenient that nine years afterwards, in 1880, there were 111 water motors driven from the low-pressure town mains, using 34,000,000 cubic feet of water annually, and paying to the Municipality 2,000l. a year. The cost of the power was not small. It was charged for at 3d. to 4d. per h.p. hour, which is equivalent to a rate of 36l. to 48l. per h.p. year for motors working continuously for 3,000 hours in the year. Since that time a new high-pressure service has been established, the water being pumped by turbines in the Rhone. On the high-pressure service the cost of the power is less. It is charged for at about 0·7d. per h.p. hour, equivalent to 8l. per h.p. year of 3,000 working hours. In 1889, the annual income from power-water sold on the low-pressure system was 2,085*l.*, and on the high-pressure system 4,500*l.* At that time the receipts for power-water were increasing at the rate of 880*l.* per annum. In 1889, the motive power distributed on the high-pressure system amounted to 1,500,000 h.p. hours, annually, there being 79 motors, aggregating 1,279 horse power.

This illustrates sufficiently the growth of the use of motive power distributed in a convenient form. The power used in pumping the ordinary water-supply for municipal and domestic purposes is not included. It will be seen later on that the works, taken as a whole, are very large and important.

The location of the windmill on the hill and the water-mill by the stream indicates how conditions of human labour have been determined by the need of mechanical energy. The earlier cotton mills were all placed where water power was available, although this had the disadvantage of taking them away from the places where skilled workmen were found and from the markets for manufactured goods. In an interesting pamphlet on the ' Rise of the Cotton Trade,' by John Kennedy, of Ardwick Hall, written in 1815, it is stated that, for some time after Arkwright's first mill was erected at Cromford, all the principal mills were built near river falls, no other power than water power having been found practically useful. After the invention of the steam-engine, manufacturing industries gathered round the coal-fields. About 1790, says Mr. Kennedy, ' Mr. Watt's steam-engine began to be understood and waterfalls became of less value. Instead of carrying the people to the power, it was found preferable to place the power amongst the people.' The tendency of the conditions created by the introduction of steam power was to concentrate the industrial population into large communities and to confine manufacturing operations to large factories. The economy of producing power on a large scale and the difficulty which then existed of distributing it to any great distance favoured the growth of the factory system. Facilities for distributing power to considerable distances are of more modern origin, and may partially reverse the tendency to concentration. Further, facility in conveying power will permit the utilisation of some sources of power hitherto not available. Waterfalls are most commonly in positions inconvenient for industrial operations. But if the power generated at the fall can be transmitted to localities where it can be conveniently used, the availability of water power is greatly increased. Mountain districts where water power is abundant may come to have an advantage over districts near coal-fields.

SOURCES OF MECHANICAL ENERGY

In these lectures motive power is treated as a commodity, producible, distributable, saleable.   The first question is as to the sources from which it can be obtained.

Wind power has been used for driving ships and mills, and now and then it is alleged that, as a source of power, wind action has been too much neglected. But its intermittence restricts its use to work which can be intermittent also.   The comparatively short periods in which the wind pressure is a considerable force make it uneconomical to attempt to do more than to utilise very moderate winds.   On the other hand, the occasional great intensity of wind action during short periods involves the necessity that structures exposed to its action should be of excessive strength and costliness.

Tidal action might, no doubt, afford an enormous amount of mechanical energy.   But, up to the present time, it has been found that the cost of embankments and machinery for utilising tidal action is so great as to prohibit its employment. The direct action of the sun's heat could be employed, but here again the cost of utilisation exceeds the value of the power obtained.   Considered practically and commercially, there are only three sources of mechanical energy of industrial importance :— (1) the muscular energy of animals ; (2) the work of water falling from a higher to a lower level and automatically restored to the higher level by the sun ; (3) the heat obtained by the combustion of fuels, transformed into mechanical energy by heat engines.

The muscular energy of animals need not be considered in these lectures. It will be convenient to consider energy derived from the combustion of fuels before considering water power.   By far the most important source of mechanical energy for industrial purposes is solid fuel burnt in the furnace of steam-boilers.   The heat produced is used to generate steam, from which mechanical energy is obtained in a steam-engine.   There are, however, disadvantages in obtaining energy from coal. First of all, at least one-fifth of the heat generated in a boiler furnace by combustion of the fuel escapes by radiation and in the chimney. Next, there is the fundamental disadvantage that, under possible temperature conditions, not more than three-eighths of the heat given to the steam (that is, three-tenths of the heat of combustion in the furnace) can be transformed in the engine into mechanical work, the remainder being necessarily rejected in the condenser. Of this 30 per cent. of the original heat energy there are further large fractions wasted, in consequence of practical imperfections of the steam engine. Cheap as coal is, the cost of the energy obtained from it is multiplied greatly by unavoidable imperfections of the processes of transformation. Further, the attendance required in operating boilers and engines, the fire and explosion risk, the nuisance of smoke and difficulty of getting rid of ashes, are all drawbacks attending the use of steam power.

Solid Fuel.—By far the most important source of mechanical energy is the combustion of various natural solid fuels, chiefly the different descriptions of coal. Coal is obtainable in very many localities, and can be transported even great distances at comparatively small cost. The solid fuels owe their thermal value almost entirely to the carbon and hydrogen they contain. It has been ascertained that the heat produced by the combustion of these constituents of fuel is as follows :—

Combustible constituents	Product of combustion	Thermal Units (Pound degrees, Fahr.) due to combustion of one pound	Weight in pounds of products of combustion
Carbon . . .	Carbonic anhydride	14,540	3·67
,, . . .	Carbonic oxide .	4,452	2·33
Carbonic oxide .	Carbonic anhydride	4,325	1·58
Hydrogen . .	Water . . .	61,530	9
,, . .	Steam . . .	52,830	9

The natural solid fuels may for practical purposes be considered to be composed of carbon, hydrogen, and a portion of incombustible impurity. In a good furnace the air supply must be sufficient to ensure perfect combustion of the carbon into carbonic anhydride, or, as the table shows, there will be a waste.

Usually the products of combustion must escape at such a temperature that the product of combustion of the hydrogen escapes as steam. Hence the amount of heat obtainable from a pound of coal containing c lb. of carbon and h lb. of hydrogen is at most

$$Q = 14540 \, c + 52830h$$
$$= 14540(c + 3.63h) \text{ Th. U.}$$

An exact determination of the calorific value of a coal can only be made by direct calorimetric test.[1] But the formula gives very approximate results for ordinary solid fuels with one small correction. In the incombustible part of the coal as actually used is included a portion of hygroscopic water, which has to be evaporated at the expense of the heat produced. Each pound of hygroscopic water absorbs in evaporation about 1000 Th. U. If a pound of coal contains w lbs. of water, then its thermal value used in a furnace from which the water escapes as steam, and neglecting at present the radiation and chimney waste, is,

$$Q = 14540(c + 3.63h) - 1000w \text{ Th. U.}$$

The following table gives the composition and thermal value of some ordinary fuels. It has been assumed that in ordinary solid fuels as burned there is at least 3 per cent. of moisture:—

Fuel	Carbon	Hydro-gen	Water	Ash.&c.	Thermal units produced per pound of fuel	Evaporation in lbs. of water per pound of fuel from and at 212°
	c	h	w			
Charcoal . . .	·85	·01	·06	·08	12.825	13·28
Wood (air dried) .	·40	·05	·20	·35	8,253	8·54
Brown coal . . .	·55	·01	·36	·08	8,164	8·45
Coke	·85	—	·03	·12	12,360	12·93
„ . . .	·91	—	·03	·06	13,230	13·69
Welsh steam coal .	·88	·04	·03	·05	14,878	15·40
„ „ „ .	·81	·05	·03	·11	14,400	14·90
Staffordshire coal .	·76	·05	·03	·16	13,670	14·14
Anthracite . .	·90	·03	·03	·04	14,650	15·17
Bituminous coal .	·70	·05	·11	·14	12,700	13·14
Poor small coal .	·50	·03	·11	·36	8,743	9·05
Illuminating gas .	·61	·26	—	·13	22,540	23·33
Petroleum . .	·85	·15	—	—	20,210	20·92
Gunpowder. . .	—	—	—	—	1,300	1·35

[1] Direct calorimetric tests are difficult, and many such tests are unsatisfactory. The best tests seem to show that the true calorific value of a fuel cannot be deduced from its analysis. The possible error in calculating the calorific value from analysis may amount to ± 6 per cent.

It will be seen that the ordinary coals do not differ greatly in heating value. The last column contains the evaporative value (at 966 Th. U. per lb. evaporated), supposing all the heat produced to be usefully expended in causing evaporation.

Liquid Fuel.—Oils which are mixtures of various hydrocarbons, and which are known as paraffin, shale oil, petroleum, or kerosene, may be obtained by the distillation of certain shales, and occur naturally in many localities. The production of oil for lighting and lubrication by distillation from shale, established first in Scotland, had assumed considerable importance when the oil wells of Pennsylvania were discovered in 1859. Since then the trade in these mineral oils has assumed enormous proportions, and new localities where such oils can be obtained have been discovered in many parts of the world. The table above shows that petroleum fuel has a high heating value. A pound of petroleum, if the heat were fully used, would evaporate 35 per cent. more water than a pound of Welsh coal. In South-East Russia petroleum refuse is largely used as fuel for locomotive and stationary boilers. Such fuel appears to have an evaporative power 25 per cent. greater than that of good coal. Mr. Holden has used some crude hydrocarbon oils in locomotives on the Great Eastern Railway. More lately refined petroleum oils have been used in internal combustion engines of the same type as the gas-engine, with very satisfactory results. The consumption of the oil in such engines is only 0·85 lb. per indicated h.p., or 1 lb. per brake h.p. per hour.

The refined oils which come to this country cost at least $3\frac{1}{2}d$. per gallon, or $\frac{1}{2}d$. a pound, or about 4*l.* 6*s.* per ton, a very high price compared with that of solid fuel. They cannot therefore be used here under boilers in producing energy on a large scale, notwithstanding their high calorific value. Used in internal furnace engines, which have a high thermal efficiency, they produce energy at a cost somewhat lower than that at which it can be produced from lighting gas, at ordinary prices, and lower than that at which it can be produced by small non-condensing steam-engines, which are thermally of poor efficiency. The crude oils cannot be transported by ship and railway safely, and are not therefore to be obtained, in this country, at any rate in large quantities.

It may, however, be noted that there is no difficulty in

transporting the crude oils by pumping through pipes, and
crude oil is now transported in that way considerable distances
in the United States, and is used in boiler furnaces on a large
scale, and apparently with economical advantage. During the
present year the whole of the boilers at the World's Fair at
Chicago, developing about 20,000 h.p., were worked with crude
petroleum, pumped direct from the wells to a store tank near the
boilers. From this the oil flowed by gravitation to the nozzles
at the boilers, and was sprayed by a steam-jet into the furnaces.
The facility of control and the absence of dirt and smoke were
very great advantages.

The same oil was supplied for working the boilers at the
Hyde Park Water Works Pumping Station at Chicago. The
author was informed that the cost of the oil was 60 cents per
barrel (42 gallons). This is equivalent to about 16s. per ton.
It was stated that, at the price at which it was supplied, the oil
was a more economical fuel for raising steam than coal at 12s.
a ton.

Gaseous Fuel.—Many of the disadvantages of solid fuel
are diminished by using the coal to produce gas, and then
generating power by burning the gas in internal combustion or
gas-engines. Gas can be transported with great convenience
in pipes, and gas-engines work with less attendance and higher
thermal efficiency than steam-engines. In transforming heat
into work, small gas-engines are enormously more efficient than
small non-condensing steam-engines and boilers. On the other
hand, ordinary lighting gas, taxed as it is with costs of distribution due to its ordinary application for lighting purposes,
is more expensive for a given calorific value than coal. The
cost of ordinary lighting gas is increased both by the need of a
large generating plant, to meet the excessive fluctuation of
demand for lighting, and by the large distributing charges
involved in supplying a very great number of small consumers.
If gas were made specially for heating and power purposes,
either coal gas of low luminous power, or water gas, or producer
gas, it could probably be distributed to power users at less than
half the present price of coal gas. Used in gas-engines, it
would then compete on nearly equal terms as regards cost with
solid coal, and would have many subordinate advantages.

M. Aimé Witz has shown by direct experiment that a gas-

engine, worked with Dowson gas, will give an effective horse
power at a total cost, including all charges for fuel, interest, and
depreciation, not greater than that at which an effective horse
power can be obtained by a good boiler and good compound
steam-engine. It is impossible to predict how far gas-engines
will replace steam-engines, but at present they have two disadvantages. They are more restricted in size than steam-
engines, and work with less efficiency at light loads.

Production of Power by Burning Town Refuse.—There is
another source of heat energy—another fuel—which is of
some importance in connection with the question of power
distribution. In addition to ordinary sewage, which is disposed
of in well-known ways, there is in large towns a quantity of
ash-bin refuse and trade refuse, which can only be got rid of
effectually and innocuously by burning. This refuse may amount
to from 100 to 400 tons per 100 inhabitants per annum. In
a steadily increasing number of towns, such refuse is now
consumed in furnaces commonly termed *destructors*. There are
thirteen towns in Lancashire and seven towns in Yorkshire
which burn refuse in destructor furnaces, and there are similar
furnaces in towns in other counties. Altogether there are about
fifty-five towns in Great Britain, in which destructors are used.
The refuse is reduced by burning to about one-third of its
weight and one-fourth of its bulk. The organic matters and
cinders in the refuse form the fuel necessary for combustion.
The residue after combustion consists of clinkers and sharp
ashes, which have a small saleable value.

The refuse burned in destructors varies greatly in composition
in different localities. An analysis of London ash-bin refuse
gave the following components : [1]—

	Per cent.
Breeze, cinders, and ashes . . .	64
Fine dust	19
Paper, straw, and organic matters . .	12
Bottles, bones, tin, crockery, &c. .	5

Even when the heat generated is wasted, it appears that
burning in destructor furnaces is the least costly method of
refuse disposal. But the amount of heat generated is not

[1] *Report on Dust Destructors by the Medical Officer and Engineer of the
London County Council*, 1893.

inconsiderable. If the heat could be utilised as satisfactorily as that of coal in generating steam, the refuse would have a calorific value equal to about one-fifth that of an equal weight of coal. To any extent to which the heat can practically be utilised there is an additional advantage in this mode of disposal. The heat of destructor furnaces is energy which, like water power, costs nothing, except for the additional machinery which must be added to the destructor furnace for the purpose of utilising it.

Various forms of destructor furnace have been tried. Five are described in a paper by Mr. Watson, of Leeds, read before the British Association in 1892. All these are to some extent modifications of the Fryer Destructor, introduced in 1876. The destructors first built were comparatively low temperature furnaces. Temperatures taken by a Siemens pyrometer, in a destructor in Whitechapel, ranged from 180° to 1,000° F. The average of eight cells was 490° F.[1] At Ealing, the temperature in the flues averaged 631° F. With this low temperature the fumes are offensive, and hence it has been necessary in many cases to have a second furnace, termed a fume cremator, in which coal or breeze is burned, and through which the products of combustion from the destructor pass on their way to the chimney. In more recently constructed furnaces, the aim is to obtain a higher temperature in the destructor furnace itself, by modifying the form of the furnace and especially by using a forced draught. In many recent destructors, the temperature in the furnace is from 1,500° to 2,000°, and this high temperature required for the inoffensive disposal of the refuse is, as will be seen presently, favourable to the utilisation of the heat generated.

Fryer's Destructor.—This has up to the present been more extensively adopted than any other. In its original form, it was a comparatively low temperature furnace. It consists of a block of 'cells' or furnace chambers internally about nine feet long by five feet wide. The cells are placed in two rows back to back, and the products of combustion pass away by a common central flue. The top of each cell is arched and the floor forms a grate, usually sloping at one in three towards the front, where there is a door for removing clinker. The refuse is charged at

[1] *Report on the Destruction of Town Refuse*. By Thomas Codrington. 1888. Also Watson, *Refuse Disposal*, British Association, 1892.

the back and, as it burns, is dragged forward along the sloping grate. Each cell will burn from eight to ten tons of refuse per day.

At Southampton a destructor with six cells of this type has been erected,[1] and arrangements are made for utilising part of the waste heat. Each cell will burn from eight to eleven tons of refuse per day. The products of combustion pass through a multitubular boiler, and thence to a chimney, 160 feet in height and 6 feet in diameter at top. There is a bye-pass through which the gases can be taken to the chimney without passing through the boiler. The steam generated is used to drive engines of $31\frac{1}{2}$ i.h.p., which compress air for working three Shone ejectors. There are also engines for preparing fodder and for electric lighting.

The garbage and ash-bin refuse is collected in covered iron tumbler carts of two cubic yards capacity. These are taken up an inclined roadway to the charging platform of the destructors. The residue after combustion consists of about 20 per cent., by weight, of the refuse burned of hard clinkers and sharp ashes. The former is used in road-making and the latter for mortar. The initial cost of the destructor, including engine-house, inclined roadway, chimney, boiler, and ironwork, was 3,723*l*. The annual cost of burning, two men being employed by day and two at night, is 251*l*. The quantity of refuse burned daily is about fifty tons. The minimum quantity burned in a day is twenty-five tons, which is sufficient to keep up steam in the boiler. This gives about seventy-five pounds of refuse per i.h.p. hour actually utilised.

Horsfall's Destructor.—This may be distinguished as a comparatively high temperature destructor, the gases escaping at a temperature of 2,000° F. The cells or furnace chambers are five feet wide. The refuse is introduced through a large hopper at the upper end. The gases escape through openings in the furnace crown, and are thence led downwards to the main flue. The grate bars are of the rocking type, with a moderate amount of motion. The grate surface is 28 square feet per cell. The clinker is removed at the front. Nine tons has been burned per cell per day. The chief peculiarity of the Horsfall furnace is a closed ashpit, with a steam jet producing an air pressure of

[1] *Proc. Inst. Mech. Engineers*, 1892.

about half an inch of water column. It is due to this that a more
active combustion is secured, with a thicker fire and with a less
air supply. Consequently, the products of combustion escape at
a higher temperature, which is better for merely destructor
purposes, and at the same time makes the escaping heat more
available for utilisation.

Modified Fryer Destructor at Leeds.—Fig. 1 shows a
destructor erected by Mr. Hewson at Leeds, which combines to
some extent the features of the Fryer and Horsfall furnaces.

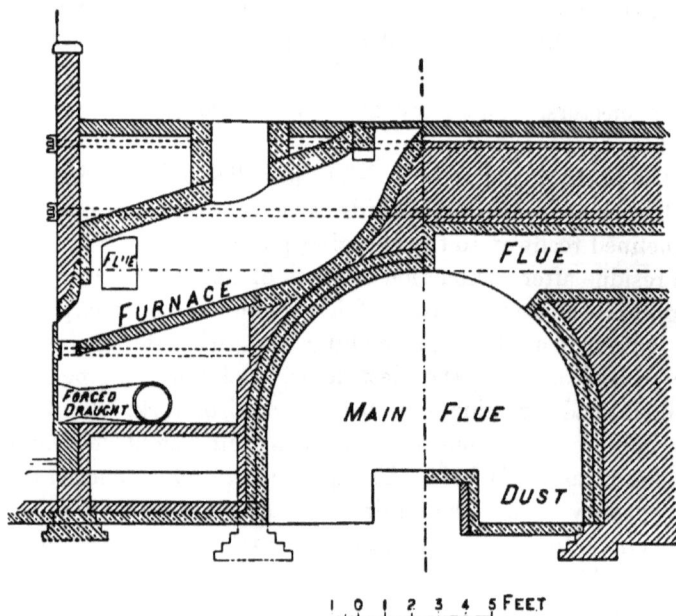

FIG. 1.

The refuse is introduced at the back, the outlet flue opening is
at the front of the furnace, and the gases are led back between
the cells to the central flue between the two rows of cells.
There is a forced draught obtained by a steam jet in a closed
ashpit.

Calorific Value of Ash-bin Refuse.—In many cases where part
of the heat from destructor furnaces has been utilised the arrangements have been imperfect, and only about four to six horses
power have been obtained per cell; that is, about a hundred-
weight of refuse burned per hour yields a horse power. This

would probably correspond to about 4 lbs. of refuse to evaporate a pound of water—a very low result.

A calculation by Mr. Watson on data obtained in working a Horsfall furnace at Leeds, in 1888, led him to conclude that a pound of refuse would evaporate about 2 lbs. of water, if the loss in radiation and chimney waste did not exceed 20 per cent. of the heat produced. Looking at the composition of refuse it does not appear that this estimate is excessive, but the data given are not very satisfactory and better data are at present wanting.

According to Mr. Keep,[1] it has been found at Birmingham that on the average 1 lb. of refuse will evaporate 1·79 lbs. of water; and at Warrington that 1 lb. of refuse will evaporate 1·47 lbs. of water.

If the heat is utilised to produce steam, by taking the gases from the destructor through a boiler, only that part of the heat which corresponds to the difference of the temperatures at which the gases enter and leave the boiler is utilised. Suppose the air enters the destructor at 60°, is raised in temperature to 1,500° in the destructor furnace, and after passing through the boiler is discharged into the chimney at 500°. Then the fraction of the heat generated utilised by the boiler is,—

$$\frac{1,500-500}{1,500-60} = \frac{1,000}{1,440}$$

or about five-sevenths. If the destructor works at 2,000°, the heat utilised in similar conditions would be,—

$$\frac{2,000-500}{2,000-60} = \frac{1,500}{1,940}$$

or about three-fourths. If the heat is to be utilised it is important to work the destructor at as high a temperature as possible.

Mr. Watson made some other experiments at Oldham. Six cells were used, burning 1½ tons of refuse per hour. The gases passed through a multitubular boiler 7 feet in diameter and 12 feet long, and the feed was measured by a meter. The temperature of the gases was 2,019° before reaching the boiler and 900° after leaving it. Thus, only about two-thirds of the

[1] *Utilisation of Town's Refuse.* C. C. Keep. British Association, 1893.

available temperature range was utilised. In two trials the
mean evaporation was found to be 2,780 lbs. per hour. Deducting
1,500 lbs. of steam used for the steam jets there was a surplus
evaporation of 1,280 lbs. per hour, or, say, 50 h.p. of energy
from six cells, burning together 1½ tons per hour. If a boiler
with larger heating surface had been used, possibly an evaporation
one-third greater would have been obtained, or say 3,700 lbs.
per hour. Deducting 1,500 lbs. used in steam jets there would
be a surplus or available steam supply of 2,200 lbs. per hour, or,
say, 88 effective h.p. from six cells burning a total of 1½ tons of
refuse per hour. This is about 40 lbs. of refuse to the effective
h.p. hour.

The chief difficulty in using the available energy of destructors
for power purposes is this. The refuse must be burned at a
nearly regular rate. But demands for power are fluctuating and
intermittent. Presently, a method of heat storage is to be
described which overcomes this difficulty. The adoption of such
a method would render the utilisation of the waste heat of
destructors much more practicable, and would give to this source
of energy a much greater importance in connection with the
problem of distributing power in towns.[1]   Of course, refuse is a
very poor fuel, and there is considerable expense in labour in
burning it. It would not, therefore, be chosen as a fuel for
raising steam. But it has to be burned for sanitary reasons.
If any profit can be made by utilising the heat, that is a gain.
The cost of the power obtained is merely the interest on the cost
of the boilers and appliances which have to be added to the
destructor, in order to utilise the heat which would otherwise be
wasted.

[1] It may be mentioned here that Prof. G. Forbes has proposed to use the
heat from destructors to work steam pumps lifting water to a reservoir on a
hill. A store of water power would thus be continuously accumulated, which
could be used to drive hydraulic motors at times when motive power was required. It will be seen later that a constant water power due to the flow of
a river is thus utilised, by pumping to a reservoir, for intermittent work, in the
systems at Zurich and Geneva. Prof. Forbes's proposal is the adaptation of
the same method to a new case.

CHAPTER II

POWER GENERATED BY STEAM ENGINES. CONDITIONS OF ECONOMY AND WASTE

SIR FREDERICK BRAMWELL, in an address to the Institution of Civil Engineers in 1885, indicated in a convenient phrase those conditions involving waste of fuel in the production of steam power, which are unavoidable when separate engines are used, but which can be diminished by central-station working. He said that we were 'every day becoming more alive to the benefit, where little power is required, or where considerable power is required intermittently, of deriving that power from a single source.' Small steam-engines are nearly always costly, uneconomical, and inconvenient. Large steam-engines and boilers working with a varying and intermittent load are in conditions unfavourable to economy. It is necessary to examine these cases in detail, and to trace the causes of waste. That is one step towards understanding in what circumstances central-station working is desirable.

Evaporative Power of Boilers at Full Load.—The following Table contains a selection of the results of the most carefully made tests of boilers. The boilers may be assumed to have been worked at nearly the full load, except where trials were made with a varying rate of evaporation. For comparison of different boilers, working with different feed and steam temperatures, the evaporation per pound of fuel is reduced to the equivalent evaporation from and at 212° F. Where possible, the influence of different qualities of coal in different trials has been eliminated by reducing the evaporation to the equivalent evaporation by a pound of pure carbon.

The first thing to notice in this Table is that the evaporation per pound of coal does not vary in different boilers so greatly as might perhaps be expected, from the great variation of construction and of the conditions of working in different experiments. Thus taking the column which gives the evaporation

TABLE I.—EVAPORATIVE EFFICIENCY

No.	Description of boiler	Authority	Indicated horses power	Steam pressure pounds per sq. in. by gauge	Coal burned per sq. ft. of grate per hour	Water evaporated per sq. ft. of total boiler heating surface per hour
					lbs.	lbs.
1	Portable (loco. type)	Kennedy & Donkin	—	61·0	12·4	3·44
2	Cornish . . .	Unwin .	256	60	7·24	1·88
3	Lancashire . .	Ellington .	179	83	20·35	3·70
4	,, . .	Donkin .	87	52	6·75	2·72
5	,, . .	Longridge.	315	89	19·82	5·72
6	,, . .	,,	338	88	22·56	6·21
7	,, . .	,,	490	84	16·0	3·32
8	,, . .	,,	493	63	15·63	3·60
9	,, . .	,,	[400]	78	17·23	4·95
10	,, . .	Donkin & Kennedy	[50]	50	10·2	2·20
11	,, . .	,,	[50]	57	18·5	4·09
12	,, . .	,,	—	79	16·0	3·79
13	,, . .	,,	[164]	77·8	16·8	4·02
14	Babcock . . .	Longridge.	[130]	98	38·0	3·54
15	,, . .	,,	[80]	61	24·7	2·34
16	,, . . .	Percy Still	214	150	14·4	3·17
17	Tubular . . .	Leavitt	252	99	7·35	2·12
18	,, . . .	,,	290	99	8·62	2·44
19	,, . .	Carpenter .	574	122	—	1·39
20	Water Tube (Thornycroft)	Kennedy .	89	182	7·74	1·24
21	,, . . .	,,	282	171	18·60	3·20
22	,, . . .	,,	449	149	29·8	4·70
23	,, . . .	,,	775	181	66·8	8·50
24	Marine . . .	Admiralty Report	[200]	—	17·27	5·94
25	,, . . .	,,	[300]	—	28·51	8·88
26	Portable Agricultural	R.A.S.E. Report	[25]	—	15·7	1·70
27	,, . . .	,,	[35]	—	20·7	3·80
	Marine					
28	Meteor . . .	Kennedy .	1,994	145	19·25	4·49
29	Fusiyama . .	,,	371	57	18·98	3·48
30	Colchester . .	,,	980	81	26·10	7·39
31	Tartar . . .	,,	1,087	144	11·93	4·13
32	Iona . . .	,,	645	165	22·4	2·73
33	Ville de Douvres .	,,	2,977	106	31·3	9·02

Figures in square brackets are approximate

OF VARIOUS TYPES OF BOILERS

Coal used	Heating value of coal per pound in Th. U.	Water evaporated from and at 212°, per pound of coal as used		Equivalent evaporation by boiler only per pound of carbon value	Remarks
		By boiler only	By feed heater		
		lbs.	lbs.	lbs.	
—	—	10·78	—	11·25	—
Welsh Steam .	14,500	11·87	—	11·87	—
Seaborne Rough	—	8·95	1·64	—	—
Small	—	11·04	—	—	—
Lancashire . .	—	8·63	1·52	—	Mean of 6 trials
" . .	—	8·09	1·13	—	Same boiler as preceding
—	—	8·53	1·23	—	Mean of 2 trials
—	—	9·48	1·35	—	Same boiler as preceding
Slack Coal . .	11,962	8·50	1·50	10·30	—
Welsh Steam . .	—	9·92	—	11·34	Perret grate
" " .	—	10·02	—	10·90	Perret grate (same boiler)
—	—	10·56	1·82	13·12	—
—	—	10·98	1·49	13·05	—
Wigan Slack . .	13,041	6·68	0·18	7·43	—
Worsley Burgy .	12,900	6·94	—	7·80	Mean of 3 trials
Welsh Merthyr .	—	12·68	—	—	—
Pocahontas . .	—	11·60	0·52	12·23	} Same boiler
" . .	—	11·35	0·48	11·98	
" . .	14,232	10·72	—	10·92	—
Welsh Steam . .	15,000	13·40	—	13·08	
" " . .	"	12·48	—	12·18	}Same boiler
" " . .	"	12·00	—	11·70	
" " . .	"	10·29	—	10·04	
Hartley Newcastle .	[15,000]	11·70	—	11·31	—
" " .	[15,000]	10·58	—	10·22	—
Welsh Steam . .	[14,500]	10·23	—	10·23	Specially skilful firing
" . .	[14,500]	9·93	—	9·93	" "
Scotch Bituminous .	12,770	8·21	—	9·62	—
Hartley Newcastle .	12,760	8·87	—	10·10	— .
Midland . . .	13,280	8·53	—	9·34	—
Welsh . . .	14,995	10·80*	—	10·44	*Corrected for moisture in steam
Walbottle Tyne .	14,830	10·63	—	10·42	—
Block Fuel . .	14,390	9·84	—	9·94	—

figures, assumed from general knowledge.

per pound of a fuel equivalent in heating value to a pound of pure carbon, the variation of evaporation for different boilers is not very great. Excluding a case or two where the boilers were certainly working badly, the lowest evaporation from and at 212° is 9½ lbs., and the highest 13 lbs. per pound of fuel. But in these boilers a square foot of heating surface produced from 1¼ to 9 lbs. of steam per hour.

Next, it may be pointed out that in boilers the evaporation per pound of fuel improves as the total quantity of steam produced diminishes, or, in other words, as the boiler load diminishes. Taking trials 15 to 18, which were made on the same boiler, or trials 19 and 20, also made on the same boiler, the results agree pretty closely with the expression

$$\text{E} = 13\cdot5 - 0\cdot4 \, w,$$

where E is the evaporation in lbs. per pound of fuel, and w the steam produced in lbs. per hour, from each square foot of heating surface. Such a rule cannot be pushed to extreme limits of working, but the general bearing of all the trials is that the efficiency of the boiler is greater as the quantity of steam produced diminishes. The boiler thus to a certain extent balances the converse action of the engine, which is less efficient for light loads.

Table II. contains the results of carefully selected steam-engine trials. In these trials it may be assumed that everything was working at its best, and that the steam and coal consumption given is the smallest realisable with the given engine, for the conditions of pressure, speed, and power stated. In these trials, also, it may be assumed that the load was constant and approximately the best for the given conditions.

Broadly, the steam consumption and fuel consumption are less for large engines than for small engines; less for quick than for slow engines; and for suitable pressures, less for compound and triple than for simple engines. Two special groups of tests have been selected to show the economy due to jacketing, and the economy due to the use of superheated steam.

In the most favourable trial conditions, as will be seen by the table, and with an economical and constant load, there is great variation in the amount of steam and coal required per indicated horse power hour.

COAL AND STEAM CONSUMPTION IN STEAM ENGINES
IN SPECIAL TRIALS

TABLE II.A.—NON-CONDENSING ENGINES

No.	Type of engine	Authority	Indicated h.p.	Piston speed	Boiler pressure	Steam per i.h.p. per hour	Coal per i.h.p. per hour	Remarks
	Simple			ft. per min.	lbs. per sq. in.	lbs.	lbs.	
1	Small Tower Spherical Engine	Unwin	7·6[1]	—	59	83·85[1]	—	593 revs. per min.
2	Semi-portable . .	,,	5	263	61	65	6·5	—
3	Horizontal, coupled slide valve	English	29·7	118	90	50·5	—	
4	,, ,,	,,	36	121	92	42·6	—	Same engine, varying load
5	,, ,,	,,	51	119	88	41·1	—	
6	Small double-acting .	Donkin	6	211	35	44	[5·0]	—
7	Horiz. McLaren . .	Unwin	3·8	248	60	46·7	—	
8	,, . .	,,	6·3	242	60	32·5	—	Same engine, varying load
9	,, . .	,,	8·2	240	60	32·86	—	
10	,, . .	,,	10·2	238	60	31·75	[3·5]	
11	Willans (central valve) slow	Willans	9	200	44	41·8	—	
12	,, ,,	,,	20	224	112	30·2	[3·4]	
13	Willans (central valve) fast	,,	16	394	36	42·8	—	Same engine
14	,, ,,	,,	26	409	74	32·6	—	
15	,, ,,	,,	34	406	122	26·0	[2·9]	
16	Beam	Hirn .	78·3	335	47	28·49	—	Super-heated steam used
17	Wheelock . . .	Hill .	140	608	96	24·9	2·5	—
18	Reynolds Corliss . .	,,	137	602	97	23·9	2·4	—
19	Harris Corliss . .	,,	134	606	96	22·0	2·2	—
	Compound							
1	Armington . . .	Meunier	84	499	117	25·3	—	Jacketed
2	Willans (central valve) slow	Willans	10	122	84	27·0	[3·0]	Not jacketed
3	,, ,,	,,	11	123	103	24·7	—	,,
4	,, ,,	,,	13	131	120	23·4	[2·6]	,,
5	Willans (central valve) fast	,,	33	403	114	21·4	—	,,
6	,, ,,	,,	36	406	135	20·4	—	,,
7	,, ,,	,,	40	401	165	19·2	[2·1]	,,
	Triple							
1	Willans (central valve) fast	Willans	39	400	172	18·5	[2·1]	Not jacketed

The figures in brackets in the last column but one are estimated on the assumption that 9 lbs. of steam are produced per pound of coal. This would be for most cases about 10·7 lbs. from and at 212° F.

[1] Brake horse-power, and steam per brake horse-power per hour.

TABLE II.B.—CONDENSING ENGINES

Stationary Engines

No.	Type of engine	Authority	Indicated h.p.	Boiler pressure	Piston speed	Steam per i.h.p. per hour	Coal per i.h.p. per hour	Remarks
				lbs. per sq. in. gauge	ft. per min.	lbs.	lbs.	
	Simple							
1	Willans, fast . . .	Willans .	7	5	380	30·00	[3·33]	
2	„ „ . . .	„	9	5	378	29·10	—	
3	„ „ . . .	„	32	50	382	25·67	—	Not jacketed
4	„ „ . . .	„	33	70	380	22·16	[2·46]	
5	Sulzer (Trois Fontaines).	Vincotte	157·5	95	433	19·84	—	
6	Corliss	Longridge	508	60	520	19·8	—	
7	„ . . .	„	488	61	520	19·3	[2·1]	
8	Sulzer . . .	Linde .	395	75	272	19·7	—	Jacketed
9	Reynolds Corliss . .	Hill .	163	96	603	19·5	1·9	—
10	Harris Corliss . .	„	166	96	606	19·4	1·9	—
11	Wheelock	„	158	96	596	19·3	1·9	—
12	Sulzer (Augsburg) . .	Linde .	291·5	90	376	19·0	—	—
13	Sulzer	„	284	87	372	18·4	[2·0]	Jacketed
	Compound							
1	Semi-portable . .	Unwin .	6	101	—	35·7	4·1	—
2	Willans, slow . .	Willans .	5·3	25	196	23·83	[2·6]	
3	„ „ . .	„	7·6	25	300	21·96	—	
4	„ „ . .	„	20	115	203	17·06	—	5 expansions
5	„ „ . .	„	25·7	100	311	16·85	[1·9]	
6	Willans, fast . . .	„	10·8	28	399	20·27	—	
7	„ „ . .	„	40	120	402	16·72	—	
8	„ „ . .	„	11·9	51	394	18·25	—	10 expansions
9	„ „ . .	„	33·2	150	397	14·82	—	
10	„ „ . .	„	13·5	150	203	16·79	—	15 to 20
11	„ „ . .	„	25	172	404	14·72	[1·6]	expansions
12	Tandem Mill . . .	Donkin .	57	53	—	20·5	1·9	—
13	„ „ . .	Longridge	862	87	442	19·8	2·15	Not jacketed
14	„ „ . .	„	888	87	442	17·8	1·78	
15	Receiver Mill . . .	„	338	95	487	17·2	2·04	-
16	„ „ . .	„	314	95	478	17·0	1·96	
17	Sulzer (Alost) . .	Vincotte	133	90	395	15·3	—	--
18	Sulzer	Soldini .	272	110	590	15·3	—	-
19	„ . . .	Sulzer .	267	90	690	14·0	[1·6]	-
20	Sulzer (Floreffe) . .	Vincotte	524	89	610	14·03	—	-
21	Sulzer (Van Hoegaarden)	„	309	88	500	13·90	—	-
22	Sulzer (Belgium) . .	„	247	85	493	13·35	[1·5]	—
	Triple							
1	Willans, slow . .	Willans .	6·7	44	302	16·99	—	
2	„ „ . .	„	23·1	170	302	12·86	[1·4]	Not jacketed
3	Willans, fast . . .	„	21·3	120	384	13·39	—	
4	„ „ . .	„	29·5	170	379	13·02	[1·4]	
5	Sulzer (Augsburg) . .	Schröter	601	145	607	12·82	—	—
6	„ „ . .	„	700	145	596	12·45	—	-—
7	Sulzer	„	198	156	460	12·2	—	...
8	Sulzer (Buda Pesth) .	Sulzer .	615	141	516	11·85	....	—-
9	Sulzer	„	360	145	—	11·70	[1·3]	—

TABLE II.B.—CONDENSING ENGINES. *Continued*

Marine Engines

No.	Type of engine	Authority	Indicated h.p.	Boiler pressure	Piston speed	Steam per i.h.p. per hour	Coal per i.h.p. per hour	Remarks
	Compound			lbs. per sq. in. gauge	ft. per min.	lbs.	lbs.	
1	Colchester . . .	Kennedy	1,979	81	520	21·73	2·90	} Not jacketed
2	Fusiyama . . .	,,	371	57	306	21·17	2·66	
3	Ville de Douvres . .	,,	2,977	106	442	20·77	2·32	
	Triple							
1	Tartar	Kennedy	1,087	144	490	19·83	1·77	} Jacketed
2	Meteor	,,	1,994	145	574	14·98	2·01	
3	Iona	,,	645	165	397	13·35	1·46	H. P. Jackd.

Pumping Engines

No.	Type of engine	Authority	Indicated h.p.	Boiler pressure	Piston speed	Steam per i.h.p. per hour	Coal per i.h.p. per hour	Remarks
	Simple			lbs. per sq. in. gauge	ft. per min.	lbs.	lbs.	
1	Beam pumping .	Mair .	123	42	223	22·0	[2·5]	} Jacketed
2	,, ,, .	,,	120	45	240	21·3	—	
	Compound							
1	Tandem pumping . .	Mair .	177	70	692	20·9	[2·3]	Not jacketed
2	Worthington [1] .	Mair and Unwin	255	60	121	17·7	1·78	
3	,, . . .	,,	296	75	124	17·4	—	} Jacketed
4	Receiver . .	Mair .	127	61	264	14·8	—	
5	Beam pumping . .	Leavitt .	290	99	241	14·2	—	
6	,, ,, . .	,,	252	99	237	13·9	[1·5]	
	Triple							
1	Worthington . . (Grand Junction)	Chadwick	182	118	–	19·3	2·25	Low duty
2	Worthington . . (Thames Ditton)	Parkes .	396	99	—		1·70	
3	Worthington [1] .	Unwin .	288	75	164	17·47	1·619	High duty
4	Worthington . . (West Middlesex)	Chadwick	260	80	164	14·10	1·66	
5	Allis pumping . (Milwaukee)	Carpenter	574	121	203	11·68	1·387	Jacketed

The figures in square brackets in the last column but one are estimated on the assumption that 9 lbs. of steam are produced per pound of coal. This would be for most cases about 10·7 lbs. from and at 212° F.

[1] On a lift of sixty feet only.

PAIRS OF EXPERIMENTS WITH AND WITHOUT JACKETS
TABLE II.C.—CONDENSING ENGINES
Stationary Engines

No.	Type of engine	Authority	Indicated h.p.	Boiler pressure	Piston speed	Steam per i.h.p. per hour	Coal per i.h.p. per hour	Remarks
				lbs. per sq. in. gauge	ft. per min.	lbs.	lbs.	
	Simple							
1	Horizontal (South Kensington)	Unwin	41·1	60·6	412	32·14	3·53	No jacket
			38 0	59·6	373	26·69	2·94	Jacket
2	Corliss (Prague)	Doerfel	146	59·2	551	22·57	—	No jacket
			159	62·2	548	19·80	—	Jacket
3	Corliss (Paisley)	Long-ridge	508	60	520	19·77	—	No jacket
			488	61	521	19·27	—	Jacket
	Compound							
1	Horizontal (South Kensington)	Unwin	44·1	66·7	343	21·06	2·30	No jackets
			45·6	67·8	352	19·52	2·13	Jackets
	Pumping Engines							
	Compound							
1	Beam pumping (Copenhagen)	Ollgaard	65·8	52·5	142 / 210	23·84	—	No jackets
			81·2	51·3	178 / 264	19·41	—	Jackets
2	Beam (Hammersmith)	Mair-Rumley	162	49·7	160·6 / 236·8	18·2	—	No jackets
			168	49·0	171·2 / 252·5	16·64	—	Jackets
	Triple							
1	Inverted pumping	Davey & Bryan	140	130	138	17·22	—	No jackets
			138	130	137·4	15·45	—	Jackets

EXPERIMENTS WITH AND WITHOUT SUPERHEATING
TABLE II.D.—CONDENSING ENGINES
Stationary Engines

No.	Type of engine	Authority	Indicated h.p.	Boiler pressure	Piston speed	Steam per i.h.p. per hour	Coal per i.h.p. per hour	Remarks
				lbs. per sq. in.	ft. per min.	lbs.	lbs.	
	Simple							
1	Beam condensing	Hirn	136	51	335	21·51	—	} Saturated steam
2	„ „	„	107	56	335	19·41	—	}
3	„ „	„	99·5	56	335	19·25	—	} Superheated steam
4	„ „	„	113	56	335	16·16	—	}
	Compound							
1	Hor. condensing (Colmar)	Unwin	475	95·7	471	19·75	3·15	Saturated steam
2	„	„	491	99	474	15·63	2·59	} Superheated steam
3	„	„	502	94·0	473	15·61	2·51	}

Taking the most favourable results which can be regarded as not exceptional, it appears that in test trials, with constant and full load, the expenditure of steam and coal is about as follows:

	Per indicated h.-p. hour		Per effective h.-p. hour	
	Coal	Steam	Coal	Steam
	lbs.	lbs.	lbs.	lbs.
Non-condensing engine . . .	2·20	20·0	2·44	22·0
Condensing engine . . .	1·50	13·5	1·75	15·8

These may be regarded as minimum values, rarely surpassed by the most efficient machinery, and only reached with very good machinery in the favourable conditions of a test trial.

It is much more difficult to get the consumption of coal by engines in ordinary daily work. What is known shows that the consumption is greater than in engine trials. Some comparatively large pumping engines, which work with a steady load night and day, and which worked with 2 lbs. of coal per effective or pump h.p. on a test trial, used 2·7 lbs. in ordinary working. The consumption was measured over many weeks, during which they were working 90 per cent. of the whole time. Here the consumption in ordinary work is 35 per cent. greater than in a test trial.

The large pumping engines of the Hydraulic Power Company are rather less favourably circumstanced for economy. They gave an i.h.p. on trial with 2·19 lbs. of coal per hour. In ordinary work they are stated to use 2·93, or about 35 per cent. more. These engines have a fairly steady load during the day and a smaller load at night.

If such a case as that of an electric lighting station is considered, where the load fluctuates very greatly, the maximum load being often four times the mean load, and the minimum load one twentieth of the mean load, then the consumption per h.p. is very much greater. Mr. Crompton has given the figures for the Kensington station, which has excellent Willans compound non-condensing engines. Those engines will work with

2 lbs. of coal per effective h.p. hour in trials at full load. The results obtained in ordinary working were as follows :

ENGINES IN ELECTRIC CENTRAL STATIONS

TABLE III.—COAL USED IN LBS. PER HOUR

	Per electrical unit generated	Per effective h.p.	Per indicated h.p.
1886	12	8·4	6·5
1890	8	5·6	4·35
1892	7	4·9	3·8

In the discussion on Mr. Crompton's paper instances were given of coal consumption at electrical stations still larger than any of these. Probably up to the present the consumption has in no case been less than 6 lbs. per unit generated; 3·8 lbs. per effective h.p.; or 3·3 lbs. per indicated h.p. This large consumption will be traced later to two classes of waste, engine waste and boiler waste, due both of them to the inefficiency caused by variation of load.[1]

In the case of small isolated motors, not generally of very good construction or well proportioned for their work, still more extravagant results have been observed. The following table gives some results obtained with small workshop engines in Birmingham :

TABLE IV.—COAL CONSUMPTION PER INDICATED H.P. HOUR IN SMALL ENGINES AT BIRMINGHAM

Nominal h.p.	Probable i.h.p. at full load	Actual average i.h.p. during the observations	Coal consumption in lbs. per i.h.p. hour during the observations
4	12	2·96	36
15	45	7·37	21·25
20	60	8·20	22·61
15	45	8·60	18·13
25	75	23·64	11·68
20	60	19·08	9·53
20	60	20	8·50

[1] The latest Board of Trade returns from electric-lighting stations confirm these figures. Taking the largest and best stations, the consumption of coal varies in different cases from 7 lbs. to 12 lbs. per unit of electricity generated. It is more than this per unit sold.

These last results are interesting both as showing how large fuel waste may be in unfavourable conditions, and also for this reason, that it is such uneconomical small engines which are displaced when central-station power distribution is introduced. It is because these small, badly-loaded engines are so extravagant that power can be distributed from a central station at a profit. As to the case of electric light stations, seeing that they are central stations of the type specially considered in these lectures, it is desirable to analyse more in detail the causes of waste.

The Chief Secondary Loss or Waste in the Action of Heat Motors.—The range of temperature, between the temperatures at which the working fluid is received by and discharged from a heat motor, is fixed by circumstances over which the engineer has little control. Thermodynamics show that for any given temperature range there is a limit to the possible efficiency of the motor. Of Q units of heat given to a heat engine, working between the temperature limits T_1 and T_2 (absolute), the part $Q \dfrac{T_1 - T_2}{T_1}$ may be converted into work, but a part which cannot be less than $Q \dfrac{T_2}{T_1}$ must be wasted. This practically unavoidable loss, arising out of the conditions under which heat and work are convertible, may be termed the primary loss in a heat engine. For steam engines, at most three-eighths of the heat given to the steam can be converted into work. In gas engines perhaps one-half might be.

Practically, no heat motor even approximately reaches so good an efficiency as this. There are secondary causes of waste, some of them important. Part of these secondary losses are due to bad construction or bad management, and can be considerably reduced by known arrangements. Part, however, are practically unavoidable, or at all events can only be partially obviated. The principal cause of the largest secondary waste of heat energy is essentially the same in steam engines and gas engines; it is a consequence of the enclosure of the working fluid in conducting metallic walls. In steam engines, water on the cylinder walls evaporates during exhaust, cooling the walls. The walls have to be re-heated during admission by the condensation of fresh steam. Nearly all the heat of the steam so condensed is wasted, during the next period of exhaust, by re-evaporation during the return stroke, when no useful work is done. The exhaust waste increases with the ratio of the area of admission surface of the cylinder to the weight of steam admitted. It increases, therefore, for light loads, because more surface is exposed during admission per pound of steam used. It is greater when the engine works slowly than when it works fast, the piston effort being the same. The evil can be diminished, but not entirely overcome, by steam-jackets, or by superheating. It is an evil specially prejudicial when engines are used which are too large for the work to be done.

In gas engines it is necessary, to prevent destruction of the cylinder by the high temperature of the burning gases, to enclose it in a water-jacket. M. Witz has shown that it is due to the cooling action of this water-jacketed wall that part of the gas is kept below the temperature of combustion. The jacket therefore diminishes the efficiency of the engine, not only by directly abstracting heat, but by preventing the full development of the gas pressure early in the stroke. As in steam engines, the evil is greater the greater the wall surface exposed at the moment of explosion. This appears to be the principal reason why initial compression of the gases is necessary for good efficiency. The gases reduced by compression to a smaller volume are exposed at the moment of ignition to a smaller area of cylinder wall.

It is useful to get a clear numerical idea of the relative importance of the cylinder wall action and the other actions during a stroke, for this cylinder wall action is the principal factor in the inefficiency due to variable load or arising out of the use of underloaded engines, both matters of importance in considering the advantages of distribution of power. Professor Dwelshauvers Dery has shown [1] how the heat exchange, between the steam and the cylinder wall during the stroke, may be represented by a diagram on the same scale as the indicator diagram. Fig. 2 shows such a diagram drawn for the data of one of Mr. Mair Rumley's engine trials. The engine was a single cylinder beam engine, with jacket, working at about $4\frac{1}{2}$ expansions, and furnishing 123 indicated h.p. The total steam used per stroke was 1·14 lb., or 31 cubic inches of water. The whole of this if condensed and spread over the cylinder wall

[1] *Investigation of the Heat Expenditure in Steam Engines*; Proc. Inst. of Civil Engineers, vol. xcviii., 1889.

would make a layer less than one-hundredth of an inch thick. The range of temperature, between the initial steam temperature and the exhaust temperature, may be taken roughly as 200°, and about 30 per cent. of the steam was condensed during admission. The whole heat of this initially condensed steam would only be sufficient to heat a very thin layer of the cylinder wall from the exhaust to the admission temperature.

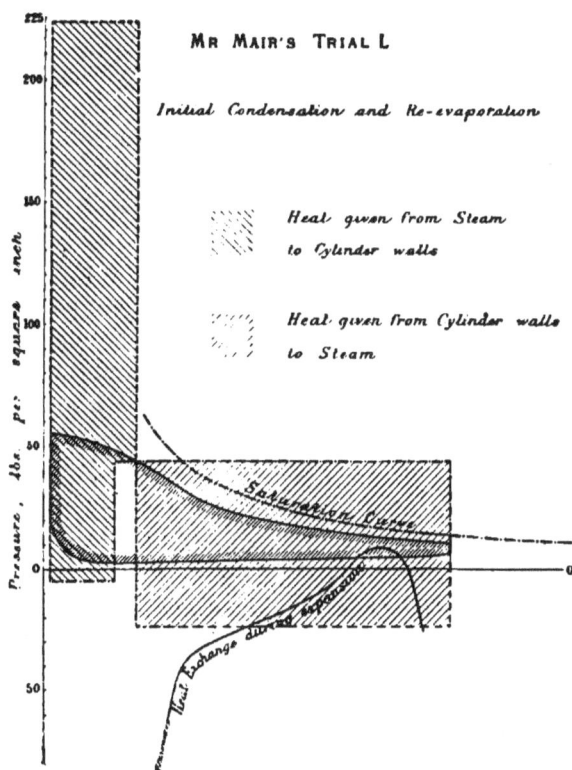

Fig. 2.

The dark shaded line in the figure is the indicator diagram of the engine. The saturation curve shows where the expansion line of the diagram should have been if there had been no condensation. The two shaded areas represent to the same scale as the indicator diagram the heat given to the cylinder wall during admission and compression, and abstracted from it during expansion and exhaust. They represent a quantity of heat first abstracted from the working fluid and finally wasted. It will

be seen easily that the cylinder wall action involves a larger
quantity of heat than the whole of the heat employed in doing
useful work.

Methods of Diminishing Cylinder Condensation.—There are
three ways in which the prejudicial action of the cylinder
wall is combated. The first is the use of a steam-jacket, the
temperature in which is higher than the mean temperature in
the cylinder. The jacket, therefore, supplies heat to the
cylinder wall during the stroke and lessens the amount of heat
which must be given by the steam to the wall during admission.
Table II.c above shows how advantageous in most cases the
use of a steam-jacket is. But though a steam-jacket reduces,
it does not prevent initial condensation. Further, the greater
the speed of the engine the less time there is for heat from the
jacket to penetrate the cylinder wall, and the less effective the
action of the jacket becomes. The second method of diminishing the cylinder wall action is to carry out the expansion of the
steam in stages; that is, to use compound or triple engines.
Then the temperature range in each cylinder and the admission
surface per pound of steam in each cylinder are less, so that to
some extent condensation is directly diminished. But also the
steam re-evaporated during exhaust in one cylinder forms part
of the admission steam to the next cylinder. Tables II. A and B
show clearly the advantages of stage expansion. There is yet
one other method of reducing the action of the cylinder wall,
and that is to use superheated steam, Table II.d. The cylinder
wall is then to a considerable extent reheated by the superheat
in the steam without condensation.

' Hirn discovered, forty years ago, that when steam is heated
above its saturation temperature, or superheated, before admission to the cylinder, the initial condensation is diminished
more effectively than by jacketing. In the years 1854 to 1865
superheating was somewhat extensively used, especially in the
Navy, and always with a marked gain of efficiency. Various
practical difficulties in the use of superheaters, especially the
danger when wrought-iron superheaters were overheated, led to
the disuse of the process.

Lately superheating has been reintroduced in Alsace, where
its advantages were first discovered, and superheaters have been
applied to a very large number of boilers. Many experiments

have been made by Alsatian engineers with saturated and super-heated steam in the same engines. In all cases they have found an economy ranging from 10 to 25 per cent. when superheated steam was used. It has been commonly alleged that the high temperature of superheated steam causes scoring or erosion of the cylinder and valve faces. Such injury did occur in the early use of superheated steam, for at that time no lubricant was obtainable capable of standing a high temperature. But this danger has probably been very greatly exaggerated. The cooling action of the cylinder is so great that superheated steam does not retain its high temperature for a sensible time after admission. With ordinary care and the use of a good lubricant, it does not appear that the engines using superheated steam suffer any injury.

Lately, the author had an opportunity of testing a large mill engine in Alsace using superheated steam. The engine was a horizontal receiver compound engine, each cylinder having four slide valves and Corliss gear. The cylinders were steam-jacketed, and were used with steam in the jackets in all the trials. The normal steam pressure was 85 lbs. to 100 lbs. per square inch. The following table gives the principal results:

TRIALS OF MILL ENGINE AT LOGELBACH WITH SATURATED AND SUPERHEATED STEAM

	Trial II. With saturated steam	Trial I. With super-heated steam	Trial I'I. With super-heated steam
Total indicated h.p. . . .	475·0	491·0	502·3
Boiler pressure, lbs. per sq. inch . .	95·72	99·05	94·0
Amount of superheating . . .	none	118°·3 F.	126°·9 F.
Pounds of steam per lb. of coal . .	6·276	6·024	6·21
Pounds of steam per i.h.p. hour . .	19·75	15·63	15·61
Pounds of coal per i h.p. hour . .	3·147	2·593	2·513
Per cent. economy of steam due to super-heating	—	20·9	20·9
Per cent. economy of coal due to super-heating	—	17·6	20·1

The coal was of poor quality.

The superheaters used in this case were constructed of cast-iron pipes of special form, under patents taken by M. Schwoerer of Colmar. The superheater is usually placed in a chamber forming part of the boiler flues; sometimes a detached super-heater with separate fire is used.

D

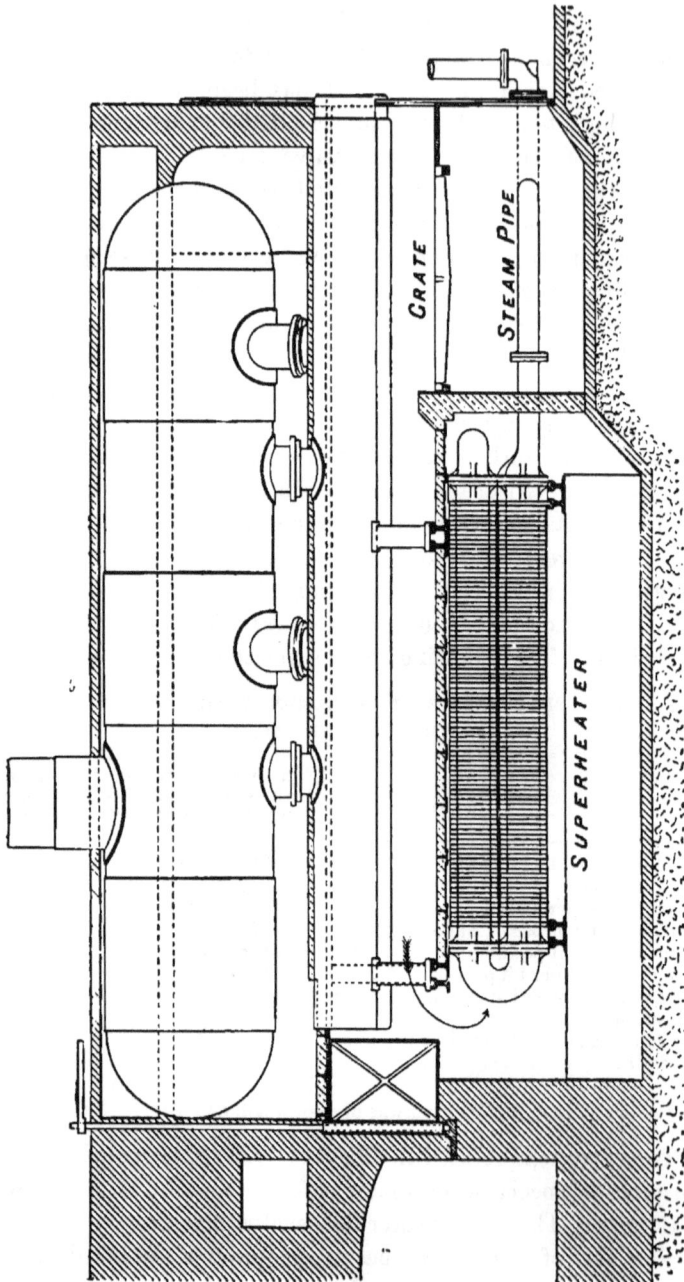

FIG. 3.

Figs. 3, 3A, show the general arrangement of a superheater on
M. Schwoerer's system. The boiler in this case is an elephant
boiler. The cast-iron coil which forms the superheater is placed
in a chamber below the
boiler, through which the
furnace gases are taken
before they are too much
cooled by contact with the
boiler surfaces. The superheater requires no attention
while the boiler is working.
Its construction is such that
there is no reasonable probability of injury or danger
from overheating.

*Load Curve and Load
Factor.*—A curve, the ab-
scissæ of which represent
time, and the ordinates
the rate of expenditure of
energy, is called a load
curve, and such curves are
very commonly drawn for a
period of twenty-four hours'
working, because, apart from
seasonal fluctuations, a day
is a natural period in the
operation of a power plant.
The ordinates may be horse-
power, or volt-ampères, or
units of any other quantity
proportional to the rate of
expenditure of energy. The
area of the curve represents
the total amount of energy
for the period considered.
A load curve may be drawn
for a single engine or

FIG. 3 A.

machine, or for a plant of many engines or machines. The load
line for a central station is that to which attention is to be

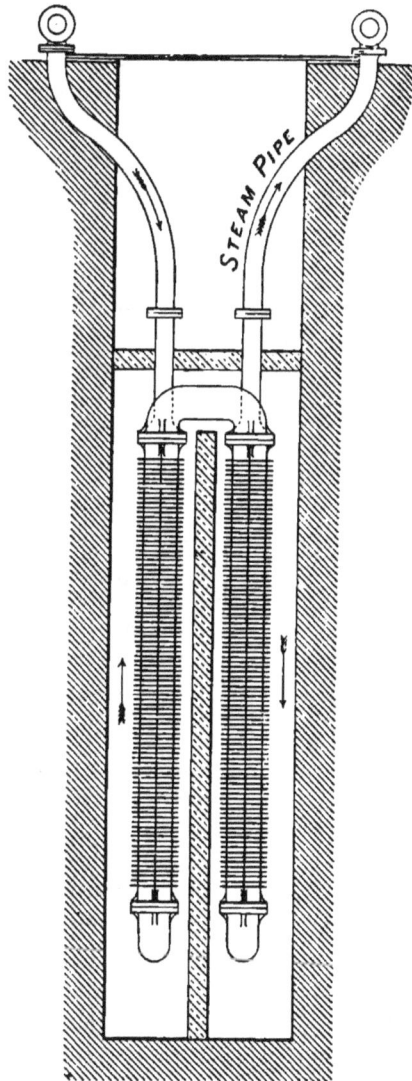

directed.   The ordinate of such a station load curve represents
the sum of the energy expended at the moment per unit of time
by all the engines or machines then in operation.

Load curves for particular cases have, no doubt, been
frequently drawn, and the influence of fluctuation in the rate of
working on economy has been noted. But it is due to Mr.
Crompton that the use of the load curve, in examining the results
of station working and in discriminating the causes of differences
in the results obtained in different stations, was first clearly
indicated.[1]

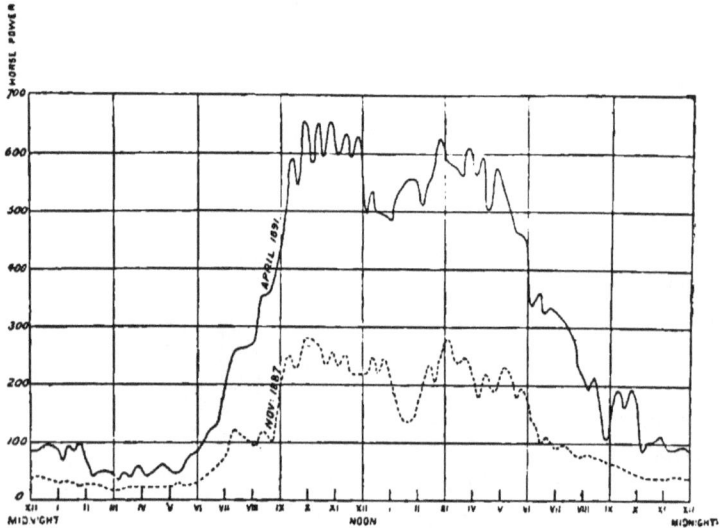

LOAD CURVES LONDON HYDRAULIC POWER STATION

1887 CURVE	*MAXIMUM = 280*	*LOAD FACTOR = ·420*
1891 CURVE	*MAXIMUM = 653*	*LOAD FACTOR = ·460*

FIG. 4.

It can be directly inferred from the examples given in Mr.
Crompton's paper that the cost of working per unit of mechanical
or electrical energy distributed, in different electric lighting
stations, depends very intimately on the form of the load curve.
Mr. Crompton introduced the term 'load factor' to express the
coefficient of fluctuation of the rate of working. There may be
various load factors, according to the precise fluctuation considered. But for the object at present in view, the consideration of the influence of variation of load on the efficiency of

[1] *Electrical Energy* ; Proc. Inst. Civil Engineers, vol. cvi.

steam plant, the load factor may be taken to be the ratio of the area of a day's load curve to the area of a rectangle enclosing it. It is equally the ratio of the average load during the day to the maximum load at any time during the day. The plant must be large enough for the maximum load. The income depends on the amount of energy delivered. The efficiency of the engines depends on the load factor. The cost of a day's working depends partly on the average output, partly on the load factor.

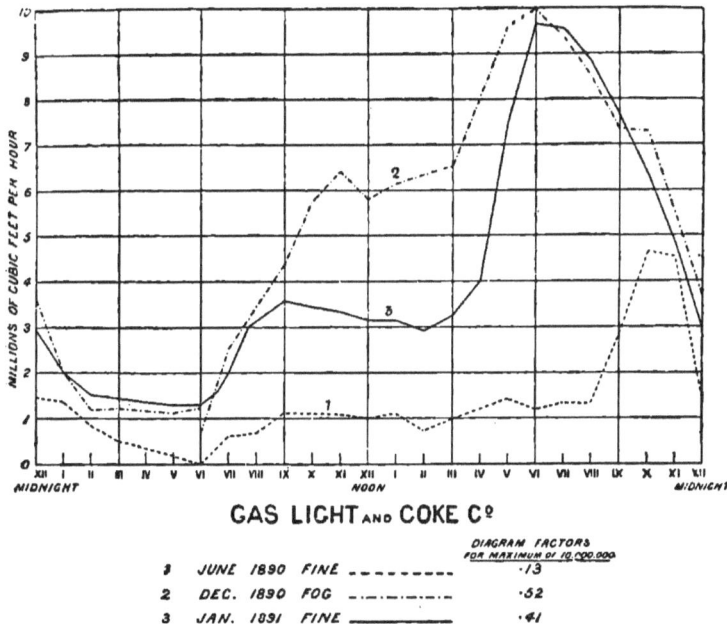

GAS LIGHT AND COKE Cº

			DIAGRAM FACTORS FOR MAXIMUM OF 10.000.000
1	JUNE 1890	FINE	·13
2	DEC. 1890	FOG	·52
3	JAN. 1891	FINE	·41

Fig. 5.

Fig. 4 gives two load curves for a day's working of one of the stations of the London Hydraulic Supply Company. These indicate the kind of fluctuation of demand which occurs in a central station supplying power for a large number of intermittently working machines, chiefly lifts and hoisting machines. Such machines are in frequent use in the day, and are little used at night. The demand for power-water pumped by the engines at the station is large and pretty constant from 9 A.M. to 5 P.M. During the remaining hours the demand is small. The load factor for the day, understood as defined above, is 0·42 in 1887 and 0·46 in 1891, when the system had been considerably extended. This shows that as the number of consumers supplied is greater, the demand is a little more uniform.

Fig. 5 shows load curves for the London Gas and Coke Company. A gas generating station is essentially a central station supplying and distributing a means of producing energy either for lighting, heating. or power purposes. The ordinates in this case represent cubic feet of gas supplied per hour. If, say, 26 cubic feet of gas per hour is assumed to be capable of

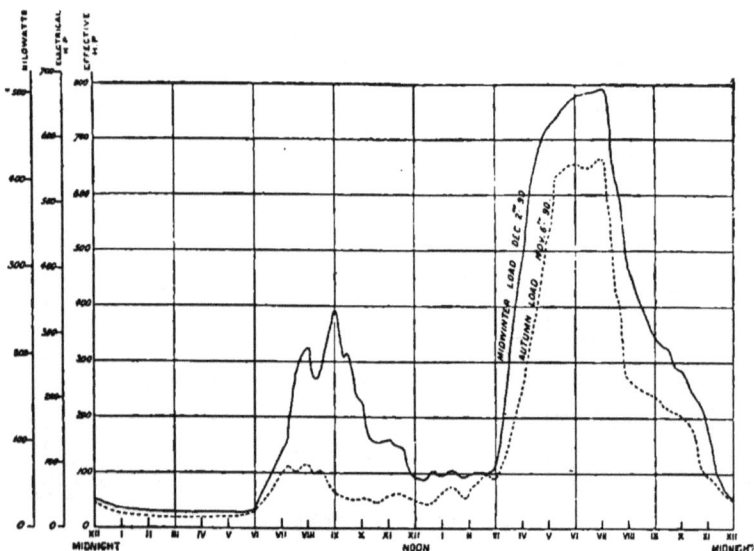

LOAD CURVES   KENSINGTON COURT ELECTRIC STATION

AUTUMN CURVE   MAXIMUM = 670   LOAD FACTOR = ·235
WINTER CURVE   MAXIMUM = 800   LOAD FACTOR = ·310

FIG. 6.

furnishing a horse power, it is easily seen that the ordinates of the curves to a suitable scale represent equally horses' power of energy supplied. In the case of the Gas and Coke Company, the largest demand is for lighting, and this is greatest in the evening. But there is also a considerable demand during the day for gas for heating and for power. The daily diagram factor was 0·41 for a day in January, falling to 0·13 for a day in June. On a foggy day in December it rose to 0·52.

Fig. 6 gives load curves for the Kensington Electric Lighting Station. As practically the whole of the electricity generated is used for lighting purposes, the period of large demand is short, and the fluctuation of demand greater than in either of the previous cases. The daily load factor is 0·24 for one of the curves, and 0·31 for the other. But for the partial use of storage batteries, the load factor would have been smaller still.

Indicated and Effective Horse Power.—In questions of power distribution it is clear that it is the effective horse power delivered at the crank shaft, and not the indicated horse power developed in the cylinder, which has to be considered. It is due to the difficulty of determining in most cases the mechanical efficiency of an engine that engineers have been content to reckon on the indicated horse power. It is true that the engine friction is not a very large fraction of the power developed, in full load trials, nor does this fraction vary very greatly at full load for different engines. But it is erroneous to assume tacitly that the engine friction is in all cases a quantity of relatively little importance, or that it is immaterial whether the steam consumption is reckoned on the indicated or the effective horse power. As the engine friction is nearly the same at all loads, then, though it is only a small fraction of the indicated power at full load, it is a large fraction at light loads. The electrical engineer who uses engine power and has exact means of measuring the quantity of power delivered to dynamos, very naturally and rightly pays more attention to effective than to indicated power.

Influence of Mechanical Efficiency on the Economy of Working with a Varying Load.—The mechanical efficiency of steam-engines, or the ratio of the effective to the indicated power at full load, is 0·8 to 0·85 for small engines, and may reach at least 0·9 for large engines. It is a little greater for non-condensing than for condensing engines, and for simple than for compound. A triple expansion engine constructed by Messrs. McLaren, tested on a brake, gave 122 i.h.p., and 107 on the brake, an efficiency of 0·88; a very good result for so complicated a machine as a triple engine necessarily is. The loss of power due to engine friction is not very great, or even for different types very variable, so long as the engines are worked at full load. It is quite otherwise, however, at light loads, and the extent to which this affects the economy of working has been overlooked.

Many experiments show that the engine friction is nearly the same at all loads.

$$\text{Let } T_e \text{ be the effective h.p.}$$
$$T_i \text{ the indicated h.p.}$$
$$F \text{ the engine friction in h.p.}$$

Then

$$T_e = T_i - F$$

and the efficiency is

$$\eta = T_e / T_i = T_e / (T_e + F)$$

The following Table gives the results of some experiments on a small non-condensing engine :

MECHANICAL EFFICIENCY OF A SMALL
NON-CONDENSING ENGINE

Indicated h.p.	Brake h.p.	Percentage of full indicated h.p.	Mechanical efficiency
12	10·16	100	·847
10·63	8·86	88·6	·833
10·156	8·34	84·6	·821
9·50	7·49	79·2	·788
8·95	7·24	74·6	·809
7·47	5·36	62·2	·718
6·273	4·791	52·3	·764
5·01	3·12	41·8	·622

The experiments were made at different times, so that there is a little irregularity in the results. The results agree approximately with the equation $T_e = 0·95\ T_i - 1·4.$

It appears, therefore, that the assumption that engine friction is independent of the load is sufficiently approximate. Suppose an engine works at 100 indicated h.p. and 85 effective h.p. at full load. Its efficiency at full load is then 0·85. At other loads it will be as follows :

TABLE V.—MECHANICAL EFFICIENCY OF ENGINES
WITH VARYING LOAD

Indicated h.p.	Effective h.p.	Efficiency
100	85	0·85
75	60	0·80
50	35	0·70
25	10	0·40
15	0	0·00

The decrease of mechanical efficiency for light loads is remarkable, and has a serious influence on the economy of working with varying load.

Careful experiments on mechanical efficiency with varying loads are not very numerous. It is useful, therefore, to give the results of some experiments on a Corliss engine of about 180 i.h.p. at full load. This engine was tried with a brake at Creusot, both condensing and non-condensing. It was found that the results agreed approximately with the following equations:

$$\text{Condensing} \qquad T_e = 0.902 \, T_i - 16$$
$$\text{Non-condensing} \quad T_e = 0.945 \, T_i - 12$$

equations which give results not differing greatly from those obtained by assuming the friction constant. The following are the calculated values of the efficiency:—

TABLE VI.—MECHANICAL EFFICIENCY OF CORLISS ENGINE
WITH VARYING LOAD

Ratio of actual effective power to power at full load	Mechanical efficiency	
	Condensing	Non-condensing
1·0	·82	·86
·75	·79	·83
·50	·74	·78
·25	·63	·67
·125	·48	·52

Influence of the Loss due to Back Pressure on the Economy of Steam-engine working.—Besides engine friction, there is another waste of energy in the steam-engine which has to an even greater extent been overlooked. The effective power is less than the indicated power by the engine friction; but the indicated power itself is less than the work done by the steam, by the amount of work done against back pressure.

In condensing engines the back pressure is comparatively small, but in non-condensing engines the back pressure exceeds 15 lbs. per square inch. Its influence on economy, even at full load, is considerable, and at light loads it may become excessively great.

In engines working, as most engines do, at constant speed, the work against back pressure is nearly independent of the load. In interpreting an indicator diagram (fig. 7) the total work done by the steam on the piston, called by some Continental writers the *absolute indicated work*, is the area $o\,a\,b\,c\,d$. The work afterwards wasted in overcoming back pressure is $o\,a\,f\,c\,d$. The difference is the effective work $a\,b\,c\,f$. The quantity of steam used depends on the absolute indicated work; the useful energy obtained on the effective indicated work. If the back pressure work is constant it becomes a larger and larger fraction of the absolute work as the load on the engine is diminished.

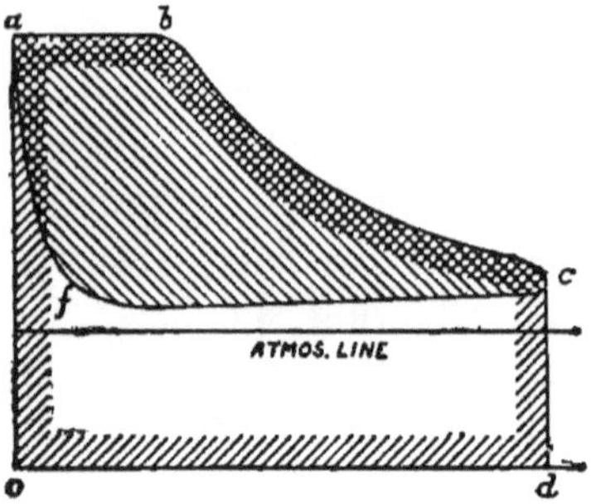

FIG. 7.

Suppose in a non-condensing engine the work against back pressure is 20 per cent. of the absolute indicated work of the steam at full load. Then for other loads the work is distributed thus :

TABLE VII.--WASTE OF WORK DUE TO BACK PRESSURE

Absolute indicated work of steam	Work against back pressure	Net or effective indicated work
h.p.	h.p.	h.p.
125	25	100
75	25	50
50	25	25
$37\frac{1}{2}$	25	$12\frac{1}{2}$
25	25	0

From the figures in the last column, the friction has to be deducted to find the useful effective work.

The following Table is calculated from the indicator diagrams of a compound engine working at nearly constant speed with a varying load. The waste work against atmospheric pressure is calculated, exclusively of the waste of work due to excess back pressure, due to resistance of passages, &c. It represents, therefore, work wasted in a non-condensing engine which is almost entirely saved in a condensing engine :

TABLE VIII.—WORK LOST IN PUMPING AGAINST THE
ATMOSPHERE

No. of trial	Effective indicated work in h.p.		Total effective indicated work	Work wasted against atmospheric pressure
	H.P. cylinder	L.P. cylinder		
	h.p.	h.p.	h.p.	h.p.
1	41·1	28·9	70·0	98·5
2	49·3	38·4	87·7	100·0
3	68·7	48·6	117·1	102·0
4	64·4	60·2	124·6	103·0
5	87·2	78·2	165·2	102·0
6	90·4	90·1	180·5	100·0

The work wasted is equal to $\frac{5}{9}$ of the effective work at full load and to $1\frac{3}{7}$ of the effective work at the lightest load.

At full load a well-designed non-condensing engine does not use much more steam per effective i.h.p. than a condensing engine. The greater back pressure in the former is partly balanced by the greater cylinder condensation in the latter, due to the greater temperature range. But with light loads it is very different. Hence condensing engines should be used, if possible, whenever the load is a very varying one.

It is stated that at the electric station at Gothenburg the fuel consumption was reduced by sixty per cent. when condensers were added to the engines. This economy was obtained in spite of the fact that at full load the engines worked nearly as economically when non-condensing as when condensing.

If the power of an engine is varied by varying its speed, instead of by varying the work per stroke, the speed being constant, then the conditions are different. If the power is varied by varying the speed, then the work against back pressure bears the same ratio to the effective work at all loads.

Influence of the Type of Engine, the Speed, and the Mode of Regulation on the Thermal Efficiency.—It has already been pointed out that there are very large thermal losses in heat engines which are not shown on the indicator diagram, and which have a very important effect on economy of working. Those thermal losses are greater also at light loads than at full load, and they vary very much with the type of engine and some other conditions of working. Unfortunately, there is not a great deal of information available as to the

steam consumption of different engines, except in the case of
full load trials. There are the experiments of the late
Mr. P. W. Willans, to which reference will be made presently.
But, except these, there are very few which afford much guidance
as to the relative economy of engines working with varying
loads.

It is possible to get a good idea of the influence of conditions
of working on the thermal efficiency in this way. Professor
Cotterill[1] has found a means of calculating the cylinder condensation in an unjacketed simple engine. The rest of the
steam used can be ascertained in other ways. By examining
the steam consumption in a variety of conditions for such an
engine, a good deal of insight may be gained applicable to all
cases. With the aid of Professor Cotterill's formula the steam
consumption has been calculated for a number of cases and
the results plotted in curves. Some check on the general
bearing of these results can then be obtained by plotting in a
similar way such experimental results as are available.

An engine has been assumed working at full load in given
conditions. Then the effect on the steam consumption of
varying the speed, the initial steam pressure, and the ratio of
expansion has been calculated. The results have been plotted
in curves which give the steam consumption in pounds per
i.h.p. hour at any fraction of full load.

$$\text{Let } \text{N} = \text{revolutions per minute}$$
$$d = \text{diameter of cylinder in feet}$$
$$r = \text{ratio of expansion}$$

Then

$$\frac{c \log_e r}{d\sqrt{\text{N}}}$$

is the condensation per pound of steam admitted to the cylinder.
Let p_1 be the initial (absolute) steam pressure and p_b the back
pressure in pounds per square inch; v_1 the volume in cubic feet
of a pound of steam at p_1. Then the indicated work done on
the piston per pound of steam not condensed is,—

$$144 p_1 v_1 \left\{ 1 + \log_e r - \frac{r\, p_b}{p_1} \right\}$$

In consequence of condensation the work per pound of steam

[1] *The Steam Engine*, 1890, chap. xi.

actually used is reduced in the ratio 1 to $1 + \dfrac{c \log_e r}{d\sqrt{N}}$. Hence, in an engine in which condensation occurs, the effective work per pound of steam is

$$144\, p_1\, v_1\, \frac{1 + \log_e r - \frac{r\, p_b}{p_1}}{1 + \dfrac{c \log_e r}{d\sqrt{N}}}$$

But a h.p. hour is 1,980,000 ft. lbs. of work. Consequently the number of pounds of steam required per i.h.p. hour will be

$$W = \frac{1,980,000 \left\{1 + \dfrac{c \log_e r}{d\sqrt{N}}\right\}}{144\, p_1 v_1 \left\{1 + \log_e r - \dfrac{r\, p_b}{p_1}\right\}}$$

For the following calculations the engine has been assumed to have a cylinder four feet in diameter, and the constant c has been taken equal to six. In considering the effect of speed, a slow engine working at $12\frac{1}{2}$ revolutions per minute, an ordinary engine working at 25 revolutions per minute, and a fast engine running at 50 revolutions per minute, have been assumed. The back pressure is taken at 3 lbs. per square inch for condensing, and at 16 lbs. per square inch for non-condensing engines.

Methods of Regulation when the Load varies.—There are three ways in which the conditions of working may be varied when the power demand varies. The speed may be varied, as is often done in the case of pumping engines. The initial steam pressure may be varied, which alters the weight of steam used per stroke by altering the density of the steam. This may be done by varying the boiler pressure or by throttling the steam. Lastly, the expansion may be varied. A case of an engine has been taken and the effect on the steam consumption of these different ways of varying the power has been calculated. Figs. 8 and 9 show by curves the results both for condensing and non-condensing engines. The results are theoretical, but they have been compared with data from various engines, and they agree with them quite closely enough for the purpose of comparison. Strictly, however, these results are applicable only to unjacketed engines.

CONDENSING ENGINES

CASE I.—*Power Varied by Varying the Speed.*—$p_1 = 110$ lbs. per sq. in.; $v_1 = 3.99$ c. ft.; normal ratio of expansion 4. The steam consumption for these conditions per i.h.p. hour is given in fig. 9. As the speed changes from 50 revolutions at full load to $6\frac{1}{4}$ revolutions at $12\frac{1}{2}$ per cent. of full load, the steam consumption rises from 18 lbs. to 25 lbs. per i.h.p. hour.

CASE II.—*The Initial Pressure Varied.*—Normal ratio of expansion 4; p_1 at full load 110. The results for a fast, ordinary, and slow engine are shown in fig. 9.

CASE III.—*Ratio of Expansion Varied.*—$p_1 = 110$. The steam consumption is calculated for a fast, an ordinary, and a slow engine. The ratio of expansion at full load is taken at 2 only. The results are given in fig. 9. It is due partly to the small expansion assumed for full load that the steam consumption per i.h.p. diminishes as the load diminishes. It would increase with small loads if different assumptions were made.

NON-CONDENSING ENGINES

The same cases with the same data are calculated and the results given in fig. 8. Only, for the case of varying expansion, the ratio of expansion at full load is taken at 4.

The results are plotted in the curves shown in figs. 8 and 9. These curves cannot be taken as absolute guides, partly because a constant value for c has been taken, partly because the formula is only trustworthy within limits. Still, they are very instructive as indicating the way in which variation of load causes a variation in the steam consumption.

Curves for Actual Engines Similar to the Theoretical Curves.— In order to test how far the curves just given agree with tests of actual engines, curves drawn in the same way for some engine trials in which the engine was tested at different loads are given in figs. 10, 11, 12. It will be seen that although these curves embrace a remarkable variety of engines, the size, the speed, the type of engine, and the initial pressure all varying considerably, yet the curves for corresponding cases are very similar to the theoretical curves previously given.

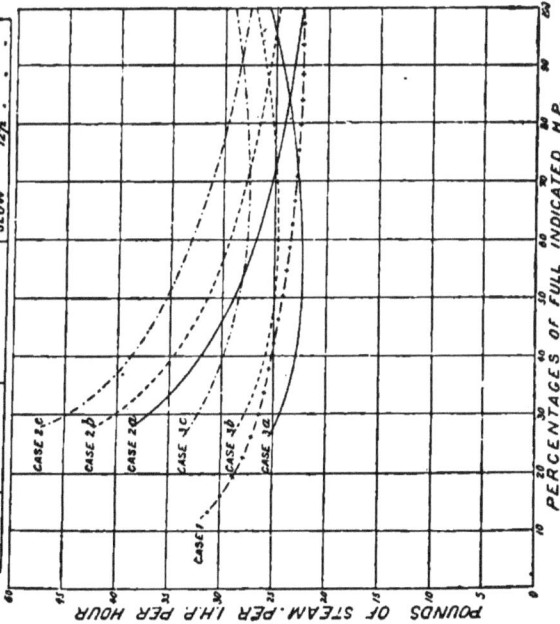

FIG. 9.

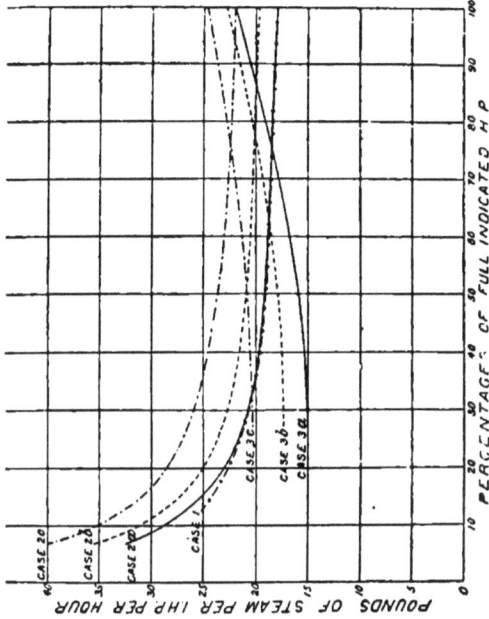

FIG. 8.

Fig. 10 shows curves drawn from data in Mr. Willans's paper on non-condensing engines, simple, compound, and triple. Also one curve for a triple condensing engine. The full curves show the effect of regulating by varying the boiler pressure or by throttling. The dotted curves show the effect of varying

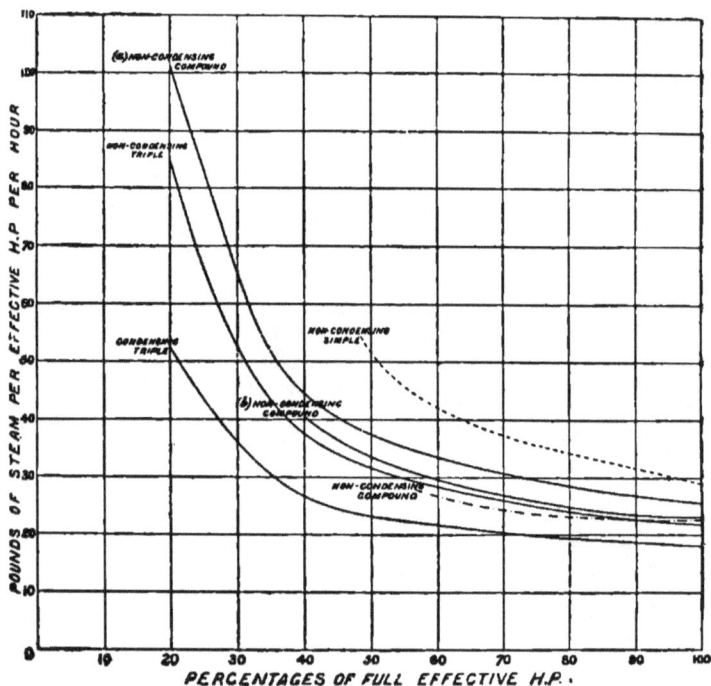

FIG. 10.

expansion, and varying both expansion and pressure. In this figure the abscissæ are percentages of the full load effective or brake h.p. Fig. 11 gives the same results, the abscissæ being percentages of the full load indicated h.p. The comparison of the two diagrams is instructive, because it shows how much more rapidly the steam consumption increases at light loads when the real useful or effective work is considered than when the indicated work is considered. It shows that to pay attention to the indicated work only is misleading.

WILLANS ENGINE

CASE	KIND OF ENGINE	VARIATION of I.H.P. by	SHEWN BY
CASE 2	NON-CONDENSING COMPOUND	VARYING BOILER PRESSURE	
"	" "	" "	
"	NON CONDENSING TRIPLE	" "	
"	CONDENSING TRIPLE	" "	
CASE 3	NON-CONDENSING COMPOUND	VARYING EXPANSION	
CASE 5	NON CONDENSING SIMPLE	VARYING P. and r	

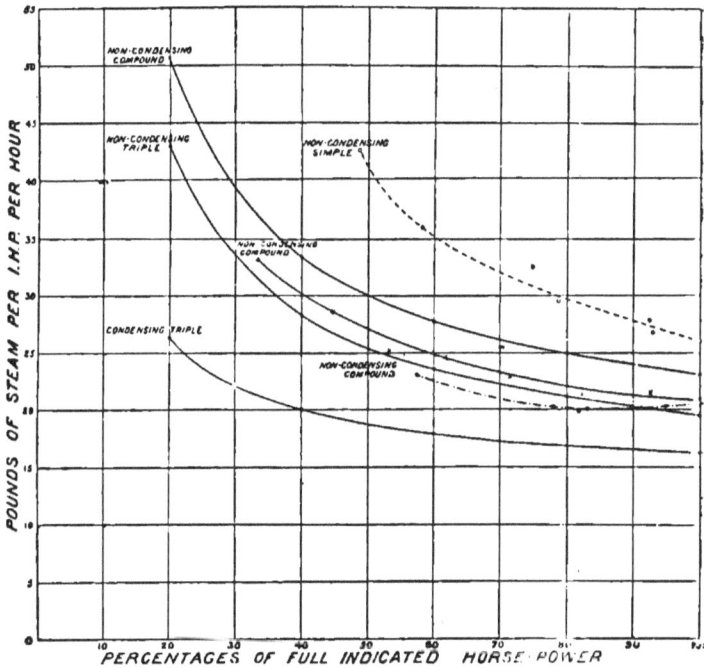

FIG. 11.

Fig. 12 shows a number of miscellaneous results selected from the few cases in which large engines have been carefully tested with varying loads. The results are more irregular than Mr. Willans's results, as might be expected, but they are instructive nevertheless.

E

Mr. Willans's Law.—In the discussion on Mr. Crompton's paper[1] the late Mr. P. W. Willans first stated a remarkable simple approximate law for the total steam consumption of an engine, working at constant speed with a constant ratio of ex-

CONDENSING ENGINES.

	SIMPLE ENGINES	VARIATION OF IHP BY	SHEWN BY
CASE 3	"MICHIGAN"	VARYING P	
	MAIR	. .	
	"GALLATIN"	. .	
	GATELEY	. .	
	COMPOUND ENGINES		
CASE 3	"BACHE"	VARYING P	
	DONKIN (NON-JACKETED	. .	
	DONKIN (JACKETED	. .	
CASE 4	"LELIA"	VARYING P, AND N.	

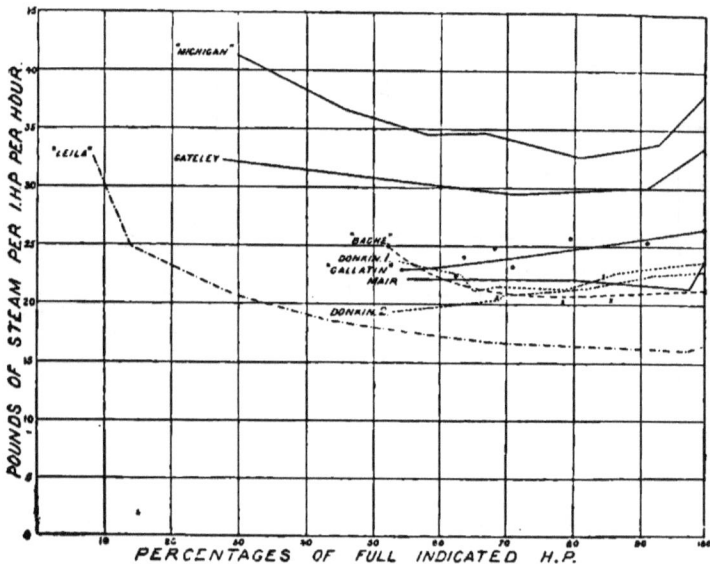

PERCENTAGES OF FULL INDICATED H.P.

POUNDS OF STEAM PER I.HP PER HOUR.

FIG. 12.

pansion. He found that the total weight w in lbs. of steam used per hour for any engine in the conditions stated was given by a linear equation of the form

$$w = a + b \text{ h.p.}$$

where h.p. is the horse-power at which the engine works. Thus,

[1] *Proc. Inst. Civil Engineers*, vol. cvi., p. 62.

for an engine of 100 indicated h.p. at full load, he obtained the following equations, in which i.h.p. is put for indicated h.p., and e.h.p. for electrical h.p., which is taken at 80 per cent. of the indicated power.

Non-condensing Triple.—(About 6·7 expansions.)

$$w = 450 + 13\text{·}75 \text{ i.h.p.}$$
$$= 725 + 13\text{·}75 \text{ e.h.p.}$$

Non-condensing Compound.—(About 4·45 expansions.)

$$w = 525 + 16\text{·}25 \text{ i.h.p.}$$
$$= 850 + 16\text{·}25 \text{ e.h.p.}$$

Condensing Triple.

$$w = 112 + 13\text{·}75 \text{ i.h.p.}$$
$$= 377 + 13\text{·}75 \text{ e.h.p.}$$

If, instead of calculating the total steam per hour w, we calculate the steam per h.p. hour, we get results which, plotted like the previous diagrams, give rectangular hyperbolas, curves which agree closely with the theoretical curves for the case of varying pressure previously given. To show the great variation of steam consumption these values are given in the following tables for full load, half load, quarter load, and one-eighth load :—

TABLE IX.—STEAM CONSUMPTION, LBS. PER I.H.P. HOUR

Indicated h.p.	Non-condensing		Condensing
	Compound	Triple	Triple
100	21·5	18·2	14·9
50	26·7	22·7	16·0
25	37·2	31·7	18·2
12½	58·2	49·8	22·7

STEAM CONSUMPTION, LBS. PER ELECTRICAL H.P. HOUR

Electrical h.p.	Non-condensing		Condensing
	Compound	Triple	Triple
80	26·9	22·8	18·5
40	37·5	31·9	23·2
20	58·7	50·0	32·6
10	101·2	86·2	51·4

Although these results are obtained from Mr. Willans's formula, they agree with remarkable exactness with his experimental results, and may be taken to be experimental results obtained with extremely good engines. Nothing could more strikingly exhibit (1) the decrease of efficiency at light loads, and (2) the very great superiority of the condensing engine at light loads.

Increase of Steam Consumption Working with a Variable Load.—Captain Sankey has applied Mr. Willans's formula to find the steam consumption of one or more engines working against a variable load, as in an electric lighting station. He takes a normal midwinter load curve and examines how the necessary current could be supplied during the twenty-four hours : (1) with an engine capable of exerting the maximum power required, (2) with smaller engines. He also considers the steam consumption when one additional engine is kept running at half speed as a stand-by in case of accident. The results, rearranged and a little modified, are given in the following table. It is assumed for convenience that the maximum load is 500 electrical h.p., and that the engines are non-condensing.

TABLE X.—STEAM CONSUMPTION IN ENGINES WORKING WITH A VARIABLE LOAD

—	Average load factor	Steam consumption in lbs. per average electrical h.p. hour	Per cent. increase of steam consumption due to variable load
I. 500 e.h.p. engine	0·22	50	108
Ia. 500 e.h.p. engine, and similar engine running at half speed	0·19	67	180
II. 200 e.h.p. engines	0·49	34·5	44
IIa. 200 e.h.p. engines, with one similar engine running at half speed	0·36	42	75
III. 100 e.h.p. engines	0·65	29·5	23
IIIa. 100 e.h.p. engines, and one similar engine running at half speed	0·53	33·1	38

Influence of Irregular Working of the Boilers on the Expenditure of Fuel.—With a varying load the steam consumption, and consequently the fuel consumption also, is increased: (1) in consequence of the decreased mechanical efficiency of the engines with light loads, (2) by the greater proportion the work expended in overcoming back pressure bears to the total work of the steam, (3) by the diminished thermal efficiency of the engine. But all these causes taken together do not explain fully the great fuel consumption in such cases as electric lighting stations. There is another very obvious cause of uneconomical working which cannot at present be estimated quantitatively for want of sufficient experimental investigation. With a very varying load boilers must be put in steam and banked up alternately, and the waste in getting up steam and allowing the boilers and brickwork to cool down again is no doubt considerable. This waste is at present unavoidable, except so far as means can be adopted to improve the load line.

Some tests made by Professor Kennedy, at the Millbank Street Station of the Westminster Electric Supply Corporation, indicate pretty clearly a boiler waste additional to the engine waste. Dividing the day into three portions, he determined the fuel consumption, the feed water evaporated, and the indicated and electric h.p. developed during each period. It will be seen that during the periods of light loading the fuel consumption per h.p. hour is very large.

TABLE XI.—COAL CONSUMPTION IN BOILERS WITH A VARIABLE LOAD

—	11 a.m. to 6 p.m. 7 hours	6 p.m. to midnight. 6 hours	Midnight to 11 a.m. 11 hours	Mean for 24 hours
Total i.h.p. hours . .	562	1,366	407	—
Total e.h.p. hours . .	400	979	260	—
Coal per i.h.p. hour lbs. .	6·88	3·26	6·26	4·65
Coal per e.h.p. hour lbs. .	9·67	4·55	9·80	6·62
Lbs. of water evaporated per lb. of coal . .	5·92	9·60	9·27	8·21
Lbs. water per i.h.p. hour .	40·7	31·3	57·7	38·3
Lbs. water per e.h.p. hour .	57·2	43·8	91·0	64·4

Perhaps the fairest way of considering the waste due to variable load will be to compare the mean consumption in the twenty-four hours with the consumption between 6 P.M. and midnight, when the load was heaviest. It will be seen that the mean steam consumption per electrical h.p. hour was 24 per cent. greater than during the period of heavy load. But the mean

consumption of coal per e.h.p. hour was 46 per cent. greater
than during the period of heavy load. The difference of
22 per cent. must be attributed to waste at the boilers, due to
irregular working. During the whole twenty-four hours the
mean evaporation in lbs. of water per lb. of coal was only
85 per cent. of the evaporation during the period of maximum
load.

*Variation of Efficiency and Fuel Consumption in Internal
Furnace or Explosion Engines.*—Gas and liquid fuel engines
receive their charge at atmospheric pressure, as well as exhausting into the atmosphere. Hence in a complete cycle
the resultant back pressure loss is comparatively small. The
engine friction, however, is rather larger than in steam engines,
and appears to be independent of the load. Hence the
mechanical efficiency decreases at light loads. Also at light
loads the combustion is in some cases less perfect, or proceeds
more slowly, and this is a cause of loss. It is well understood
that gas and petroleum engines should be worked as far as
possible at full load. At the Dessau Electric Station, which is
worked with gas engines, large secondary batteries are used to
store the surplus energy when not required for supply. and to
obviate the necessity of working the engines at light load. On
the other hand, engines of this type have the very great
advantage that they can be started in a few minutes when
required, and stopped whenever they are not wanted. There is,
in a station worked with such engines, no loss like that due to
irregular working of the boilers.

From some experiments on gas and petroleum engines, the
author obtained the following approximate equations for the
amount of fuel used.

Let w = fuel used per brake h.p. hour.

p = fraction of full load at which the engine is working.

For gas engines using lighting gas

$$w = \frac{6 \cdot 4}{p} + 15 \cdot 25 \text{ cubic feet of gas per hour.}$$

For petroleum engines

$$w = \frac{0 \cdot 4}{p} + 0 \cdot 6 \text{ lbs. of oil per hour.}$$

TABLE XII.—FUEL CONSUMPTION IN INTERNAL
FURNACE ENGINES

Brake load in per cent. of load at full power	Gas engine. c. ft. of gas per brake h.p. hour	Oil engine. lbs. of oil per brake h.p. hour
100	21·65	1·00
75	23·78	1·13
50	28·05	1·40
25	40·85	2·20
12½	66·45	3·80

It will be seen that the cost in fuel per h.p. hour increases greatly at light loads.

The Steam Turbine.—It is very many years since Girard predicted that ultimately steam turbines would supersede reciprocating engines. Although many engineers have experimented on steam turbines, Girard's prediction is still unfulfilled. But important progress has been made in the use of steam turbines. The electric stations at Cambridge and Newcastle are worked with these motors, and the steam turbine may have an important part to play in central station working.

Credit is due to Mr. C. A. Parsons for first completely facing the mechanical difficulties involved in designing a steam turbine. Of these the most obvious is the excessively great number of rotations at which any steam turbine must run. The circumferential speed of any turbine must depend on the velocity due to the pressure head of the fluid used. Taking steam of 140 lbs. absolute pressure, the weight of a cubic foot is 0·3148 lbs.; the height due to the pressure is $(140 \times 144)/0·3148 = 64,000$ feet; the velocity due to this height is roughly $8·02\sqrt{64,000} = 2,030$ feet per second. It is at some large fraction of this enormous velocity that the circumference of a steam turbine must run if any good efficiency is to be obtained.

Mr. Parsons partially met the difficulty by using multiple turbines, so as to divide the pressure difference to be dealt with into stages. In his earliest form of turbine, a series of axial flow turbines were placed on a common shaft, thirty or forty in number, with fixed guide vanes between each pair of turbine rings. Several hundred of these axial flow multiple turbines were constructed for electric light work, the aggregate horse power amounting to 4,000. The speeds of rotation were 10,000 revolutions per minute and upwards. Mr. Parsons has now modified the construction of his turbine and made it a radial outward flow turbine. The turbines in this type are arranged in concentric rings on a rotating disc, with fixed guide vanes between each pair of rings. The steam works its way radially through a series of guide vanes and wheels outwards, and then passes to the centre again to flow through another series. Professor Ewing, F.R.S., has tested one of these turbines of 135 electrical h.p. The consumption of steam was 27·6 lbs. per e.h.p. This corresponds to a reciprocating engine working with 20 lbs. of steam per i.h.p. Since then still better results have been obtained, and the remarkable feature is shown that the increase of steam consumption per h.p. with low loads is much less than with reciprocating engines.

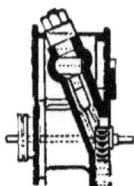

FIG. 13.

FIG. 14.

Another steam turbine, less generally known at present, is one invented by M. Gustav de Laval. Unlike that already described, this is a simple impulse turbine. The steam is discharged through two, four, or more nozzles, acquiring in an expanding mouthpiece the full velocity due to the pressure head, and issuing at atmospheric pressure with its energy converted into kinetic energy. It is received on the vanes of a single simple impulse turbine passing through it axially. Fig. 13 shows one of the nozzles discharging into the wheel. Fig. 14 shows the turbine wheel (5), on its shaft (4), and at (3) the pinions of a pair of spiral gears, by which the speed is directly reduced in the ratio 10 to 1. The complete expansion of the steam to atmospheric pressure in the nozzle (or to a lower pressure if an ejector-condenser is used) is determined by the proportions of the nozzle. The velocity of discharge for steam of 75 lbs. pressure expanding to atmospheric pressure is given as 2,625 feet per second; expanding to a vacuum pressure of $1\frac{1}{2}$ lbs. per square inch, at 4,600 feet per second. The speed of the turbine is limited by the resistance to centrifugal tension. In a 5 h.p. turbine the peripheral speed is stated to be 574 feet per second, and the number of revolutions 30,000 per minute. One great difficulty with turbines running at so high a speed is to ensure dynamical balance. In the Laval turbine the shaft is very small in diameter, and has a considerable unsupported length. If the balance is not perfect the shaft bends till the axis of rotation becomes an axis of inertia. There are then no vibrations. The spiral gearing takes up the end thrust and also runs perfectly quietly. The speed governor is very small, simple, and effective.

Experiments have been made on a 50 h.p. Laval turbine by Professor Cederblom. The steam pressure by gauge varied from 108 to 122 lbs. per square inch. A Korting ejector-condenser was used, and the pressure in the exhaust pipe of the turbine was 1·7 lbs. per square inch. The turbine developed 63·7 brake h.p. with a steam consumption of 19·73 lbs. per brake h.p. hour, and a consumption of 2·67 lbs. of coal per brake h.p. hour. These are extremely remarkable results.

The important feature of such a turbine, besides its economy of working, is its extreme simplicity and the absence of wearing parts. It can be regulated for light loads by stopping the flow through one or more of the nozzles. It ought, therefore, to have an efficiency with light loads little less than that with full load.

CHAPTER III

THE COST OF STEAM POWER

THE probable cost of steam power in any given case can only be determined by careful estimates in which local conditions are taken into account. The cost of coal, facilities for obtaining water, the cost of labour, even the type of engine and character of the buildings required are more or less different in different cases. Further, the way in which the power is applied, the number of hours the engine is used per day, and the regularity of the load during working hours affect very much the cost. Certain typical cases may, however, be taken, and an average estimate made of the cost in such cases. The case will be taken first of engines used in industry and working a regular number of hours daily with a nearly regular load. This will afford some indication as to how far motive power, supplied from central stations by some method of transmission, can be used economically, in place of power generated locally by steam engines. Then the special case of the cost of power, generated by steam in central stations for distribution will be considered.

Cost of Engines, Boilers, and Buildings.—With engines of 100 h.p. or more the cost can be pretty definitely stated, and the total cost of engines and boilers per h.p. does not vary very greatly with the type of engine adopted. For if a cheaper and simpler type of engine is selected, then, its efficiency being less, the boilers have to be larger. But with small engines the cost per h.p. increases very considerably, both because small engines are less efficient and because they are more expensive to construct.

It will be assumed for the following estimates that the total cost erected of engines and boilers, with pipes and auxiliary apparatus and such buildings as are necessary, may be taken to be as follows :—

TABLE XIII.—COST OF STEAM PLANT

Indicated horse-power	1	10	50	200
Effective h.p.	·7	7·5	40	165
Cost per i.h.p. in £ . . .	56	30	24	20
Cost per effective h.p. in £ .	80	40	30	25

In determining the annual cost interest will be taken at 5 per cent., and maintenance (repairs) and depreciation at $7\frac{1}{2}$ per cent. *Cost of Coal and Petty Stores.*—In the following estimates coal will be taken at 20s. per ton. The amount of coal required must be calculated so as to allow for lighting up boiler furnaces, for waste due to cooling of boilers and brickwork when steam is let down, and for working auxiliary apparatus such as feed pumps.

TABLE XIV.—WORKING COST OF STEAM PLANT

Indicated horse-power	1	10	50	200
Effective h.p.	·7	7·5	40	165
Coal per i.h.p. hour, lbs. . . .	8	$5\frac{1}{4}$	$2\frac{3}{4}$	2·0
Coal per effective h.p. hour, lbs. . .	$11\frac{1}{2}$	7	$3\frac{1}{2}$	$2\frac{1}{4}$

The cost of petty stores will be taken at 0·25*l.* per effective h.p. per annum in the case of moderately large engines working ten hours a day. In other cases a proportionate estimate will be made. *Cost of Labour.*—For driving, stoking, and cleaning; an allowance of 1·2*l.* per annum per effective h.p. for 3,000 hours, or 0·6*l.* per annum for 1,000 hours, will be made. In the case of engines of 10 h.p. or less, however, the labour reckoned on the h.p. costs considerably more.

TABLE XV.—COST OF AN EFFECTIVE H.P. PER YEAR OF 1,000 WORKING HOURS. THE ENGINE WORKING REGULARLY WITH NEARLY FULL LOAD

Indicated horse-power of engine	1	10	50	200
	£	£	£	£
Interest at 5 per cent. on engines, boilers and buildings	4·00	2·00	1·50	1 25
Maintenance and depreciation at $7\frac{1}{2}$ per cent.	8·00	3·00	2·25	1·88
Coal at 20s. per ton	5·13	3·12	1·56	1·01
Petty stores	0·50	0·30	0·20	0·15
Labour	6·25	3·00	0·80	0·70
Total cost of 1 effective h.p. per year of 1,000 hours in £	23·88	11·42	6·31	4·99
Cost in pence per effective h.p. hour .	5·75	2·84	1·51	1·20

TABLE XVI.—COST OF AN EFFECTIVE H.P. PER YEAR OF 3,000 WORK-ING HOURS. THE ENGINE WORKING REGULARLY WITH NEARLY FULL LOAD

Indicated horse-power	1	10	50	200
	£	£	£	£
Interest at 5 per cent. on engines, boilers and buildings	4·00	2·00	1·50	1·25
Maintenance and depreciation at 7½ per cent.	6·00	3·00	2·25	1·88
Coal at 20s. per ton	15·39	9·36	4·68	3·03
Petty stores	0·75	0·45	0·30	0·25
Labour	12·50	6·00	1·50	1·20
Total cost of an effective h.p. per year of 3,000 working hours in £	38·64	20·81	10·23	7·61
Cost of an effective h.p. hour in pence . .	3·10	1·66	·82	·61

The results given in these tables are plotted in figs. 15 and 16. The extremely rapid increase in the cost of working for small powers is very striking.

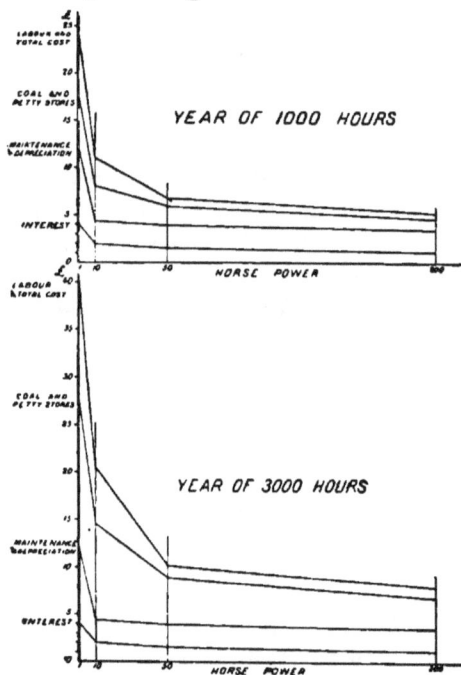

FIGS. 15 AND 16.

Annual Cost of an Effective H.P. per Annum obtained by an Engine working with Dowson Gas.—It will be useful to place alongside these estimates of the cost of a steam h.p. an estimate of the cost of a h.p. obtained from a gas engine. For comparison the very careful estimate of Professor Witz may be taken based on experimental trials of an engine of 112 i.h.p. or 77 effective h.p. worked with Dowson gas. The total cost of the engine with pump and pipes was 944*l.* or 8·2*l.* per i.h.p. The gas generator cost 280*l.*, or 2·5*l.* per i.h.p. Foundations and erection (without buildings) cost 68*l.*, or 0·61*l.* per i.h.p. The total cost without buildings was therefore 11·3*l.* per indicated or 17·2*l.* per effective h.p., a cost about equal to that of a steam engine plant of the same power. Professor Witz takes the cost of anthracite at 25*s.* a ton and coke at 28*s.* a ton. He allows for interest and depreciation 15 per cent. The gas consumption is taken at 84 c. ft. per effective h.p. hour, which allows nothing for irregular working. Professor Witz's figures are reduced to a year of 3,000 working hours.

TABLE XVII.—COST OF AN EFFECTIVE H.P. IN A GAS ENGINE USING DOWSON GAS, PER YEAR OF 3,000 WORKING HOURS

	£
Interest and depreciation at 15 per cent.	2·78
Anthracite and coke	2·36
Petty stores	·40
Wages	·96
	6·50

The cost appears to be slightly less than that of a steam engine of corresponding power. The cost is equivalent to 0·51*d.* per effective h.p. hour.

Cost of Steam Power in Central Stations.—The case of a central station worked by steam power differs from those previously considered, in consequence of the excess of plant required and the waste due to working against a varying load.

In such a generating station, whether supplying electricity or energy in any other form, it is usually necessary to work night and day. Part of the engines must work 8,760 hours in the year, but for a large fraction of the time much of the plant is standing idle. The demand for motive power purposes is greatest during the day hours, that for lighting during the evening hours; during part of the night the demand for any purpose is very small. It follows that the plant required must be much larger than that which would be required to meet the average demand, if that could be supplied uniformly during the

24 hours. Further, there must be a reserve of power, so that any engine or boiler can be laid aside for examination or repair without hindering the work of the station. That reserve will be taken to be 25 per cent. of the whole power.

The earning power of the plant depends on the average demand and average rate of working. The coal and labour depend on this also, but are increased in consequence of the uneconomical conditions of working. The interest and depreciation must be calculated on the maximum output of which the plant is capable.

How enormously the cost per h.p. per annum may be increased by the conditions which arise in central station working will be shown by an examination of the returns of cost at some electric lighting stations.

Cost of Engines, Boilers, and Buildings reckoned per Indicated H.P. at Full Load.—The following data will be assumed in the following calculations of the cost of central station working :—

TABLE XVIII.—COST ASSUMED FOR STEAM PLANT

	Cost in £ per i.h.p. at full load	
Condensing compound engine, with the necessary pipes, pumps, &c., for supplying condensing water	7·5	
Erection of engines	0·3	
Foundations	1·1	
	—	8·9
Boilers, steam pipes and auxiliary engines	4·3	
Erection of boilers	0·5	
	—	4·8
Buildings :—		
Engine house, boiler house, and coal store	5·0	
Chimney	1·0	
		6·0
Total		19·7

High speed condensing engines may be obtained, erected, but exclusive of boilers and buildings, as low as 5*l.* per i.h.p. On the other hand, there are large engines costing 30*l.* per i.h.p. The cost varies with the piston speed.

Working Cost of Engines reckoned also at per I.H.P. at Full Load.—The cost of coal will be taken at 7*s.* per ton. The cost of oil and petty stores will be taken at 0·28 pounds per i.h.p. per year. Labour is a difficult item to estimate, because it depends so much on management and conditions of working. The cost of labour will be taken at 2l. per i.h.p. actually exerted per annum.

The following rates will be assumed for interest and depreciation :—

		Per cent.
Interest on capital cost of plant	. . .	4
Maintenance and depreciation :—		
Buildings	.	2
Machinery	.	$7\frac{1}{2}$

From what has been said it will be seen that the annual cost of a h.p. depends on the distribution throughout the day of the work to be done. If the work is regular and the engine works at nearly full load the cost of the h.p. is comparatively small. On the other hand, if the work is very irregular, larger engines are required, the working is inefficient, and the cost is comparatively large. Two limiting cases will be considered.

CASE 1.—*Conditions similar to those of an Engine pumping to Reservoirs, night and day, all the year round.*—Such an engine may be taken to work 90 per cent. of the whole year, or 7,884 hours in the year. For one effective h.p. of work done there must be exerted $1/0.85 = 1.176$ i.h.p., allowing for engine friction. And for every 1.176 i.h.p., engines of 1.47 i.h.p. must be provided, to allow the necessary reserve.

CASE 2.—*Engines working in Conditions similar to those of an Electric Lighting Station.*—The engines work all through the year, but the maximum demand is four times the average demand. For every effective h.p. the engines must exert (neglecting the variations of mechanical efficiency) 1.176 i.h.p., and for every 1.176 i.h.p. of average demand there must be provided engines capable of exerting 4.70 i.h.p. during hours of maximum demand. Further, to allow a reserve the engine power in the station must be 5.87 i.h.p. for every effective h.p. of average demand.

CASE 1.—*Engines working on a very Regular Load, in Conditions similar to those of an Engine pumping to a Reservoir.*—Here for one effective h.p. exerted during 7,884 hours annually, engines of 1.47 i.h.p. must be provided. Such engines may be taken to use 14 lbs. of steam per i.h.p. hour in test trials. But in ordinary work $7\frac{1}{2}$ per cent. more must be allowed for leakage, working auxiliary engines, and less careful attention. This makes the consumption 15 lbs. of steam per i.h.p., or $15 \times 1\cdot176$ $=18$ lbs. per effective h.p. hour. At 9 lbs. of steam per lb. of coal, allowing also 5 per cent. for lighting and banking fires, the engine would use $2\cdot1$ lbs. of coal per effective h.p. hour. There are very few engines working with quite so low a consumption.

TABLE XIX.—COST OF INSTALLATION PER E.H.P.

	£
Cost of engines for 1 effective h.p., or with reserve 1·47 i.h.p. = $1\cdot47 \times 8\cdot9$.	13·3
Cost of boilers = $1\cdot47 \times 4\cdot8$	7·0
Cost of buildings = $1\cdot47 \times 6\cdot0$.	8·8
Total	29·1

ANNUAL COST OF WORKING PER EFFECTIVE HORSE POWER

	£
Interest at 4 per cent.	1·164
Maintenance and depreciation : —	
Buildings at 2 per cent..	0·176
Machinery at $7\frac{1}{2}$ per cent.	1·522
Total fixed annual cost	2·862
Coal, 2·1 lbs. for 7,884 hours, at 7s. per ton	2·587
Petty stores	0·327
Driving, stoking and cleaning .	2·334
Total annual cost .	8·110

This is equivalent to $0\cdot24d$. per effective h.p. hour.

CASE 2.—*Engines working with very Variable Load in Conditions similar to those of an Electric Lighting Station.*—Here, for one effective h.p. supplied on the average throughout the year, engines of $5\cdot87$ i.h.p. have to be provided. On account of the inefficiency and waste, due to the variation of the load, it is best to estimate the steam and coal from experience in similar cases. Probably no electric lighting station at present works with quite so low a consumption as 6 lbs. of coal per hour per electrical unit supplied. A consumption of 9 lbs. is probably much more common in the best managed stations. Six lbs. of coal per electrical unit corresponds to $3\cdot8$ lbs. per effective h.p. hour.

TABLE XX.—COST OF INSTALLATION.

	£
Cost of engines for one average effective h.p., with reserve 5·87 i.h.p. = 5·87 × 8·9	52·2
Cost of boilers = 5·87 × 4·8	28·5
Cost of buildings = 5·87 × 6·0	35·2
Total cost .	115·9

ANNUAL COST OF WORKING PER EFFECTIVE H.P.

	£
Interest at 4 per cent.	4·64
Maintenance and depreciation :—	
Machinery at 7½ per cent.	6·05
Buildings at 2 per cent..	0·70
Total fixed annual cost .	11·39
Coal, 3·8 lbs. for 8,760 hours, at 7s. per ton	5·20
Petty stores	0·33
Driving, stoking and cleaning .	2·33
Total annual cost	19·25

This is equivalent to 0·51d. per effective h.p. exerted on the average throughout the year.

Cost of a H.P. at existing Electric Lighting Stations.—It is perhaps not entirely fair to take the cost of working of electric lighting stations as a guide to the cost of steam power. They have been recently established, they work under difficult conditions, and the best methods of economising cost have probably not yet been arrived at. On the other hand, they are central stations of the kind discussed in these lectures, and accounts of the cost of working are published in returns made to the Board of Trade.

To be as fair as possible to electrical engineers the case of Bradford may be taken, where according to the returns a unit of electricity supplied is generated more cheaply than at any other station. In dealing with the figures in the returns, the charges under the heading 'salaries of manager, engineer, &c.,' and those under the heading 'redemption fund,' are discarded. Further, half the cost under the headings 'depreciation' and 'repairs and maintenance' is also subtracted, because under these headings are included charges not belonging to the cost of generating power. It would not make much difference if a larger or smaller fraction had been subtracted. After making these deductions, the cost of a unit of electricity supplied at

F

Bradford, mainly, if not entirely, attributable to the cost of producing power, is 2·1d. Now, the mechanical value of an electric unit is 1·34 h.p. hours. Taking the average efficiency of the dynamo at 0·85, then one unit corresponds to 1·34/0·85 = 1·57 effective h.p. of the engine. Calculated on this basis it appears that the cost of an effective h.p. per year of 8,760 hours at Bradford is 49l. The cost of coal and petty stores alone, exclusive of all charges for labour, interest, and depreciation, is 14·6l. At most other stations for which returns are made, the cost reckoned in the same way is considerably greater.

CHAPTER IV

THE STORAGE OF ENERGY

IN most applications of the energy derived from fuel the fluctuation of the demand for power involves increase of cost in two ways : (1) The generator of energy must be large enough to meet the maximum demand, so that its cost for a given output is greater than it would be if the demand for power were constant. (2) In the working of such a large generator there is waste of fuel and increased cost of superintendence. In steam central stations which must supply energy throughout the twenty-four hours of the day the demand for power fluctuates very greatly, and the increased cost of power due to this is serious. Probably electrical central stations for lighting are those in which the fluctuation is greatest of all, and hence electrical engineers more than others have sought means of storing energy in times of small demand, to be used in times of large demand.

To completely meet a varying demand by generators of energy worked at a uniform rate, there must be an amount of storage of energy satisfying two conditions. Twenty-four hours may be taken as the natural period of the fluctuation of demand for energy. In each twenty-four hours there will usually be two periods, one in which the demand falls below the average demand, and one in which the demand exceeds the average demand. The excess of energy to be supplied during one period will, on the average, be equal to the deficiency in the other period. If the generators are worked at a uniform rate, then all the energy supplied in excess of the mean demand in one period must be taken from storage, and must have been put into store during the period in which the demand fell below the mean demand. But this is

not the only condition to be satisfied. With some kinds of
storage the rate at which energy can be taken out of store
is unlimited. In other cases it is limited, and then the
storage must be so arranged that the rate at which energy
can be taken out of store is equal to the difference between
the maximum rate at which energy is required and the mean
rate.

Fly-wheel Storage.—All steam engines are provided with
fly-wheels which store and re-store part of the energy generated
in the cylinders. Let w be the weight of a fly-wheel in
tons, v its velocity of rim in feet per second; then the
total kinetic energy is (approximately enough for the present
purpose) 2,240 w $(v^2/2g)$. For a range of variation of velocity
from v_1 to v_2 the amount of energy alternately stored and re-
stored is

$$2,240 \text{ w } \frac{v_2^2 - v_1^2}{2g} \text{ foot lbs.}$$

Suppose a fly-wheel weighs twenty tons, its rim velocity
cannot generally exceed fifty feet per second, for reasons of
strength. If a 5 per cent. variation of speed is permitted (which
is more than is usually allowed), the amount of energy
alternately stored and re-stored will be 169,800 foot lbs.,
or 0·086 h.p. hour, an insignificant quantity compared with
the work which would be done by an engine with such a
fly-wheel. If, however, the fluctuation of demand for energy
occurred in half a minute, the fly-wheel would supply in that
time 10·28 h.p., which might have a very useful effect in
diminishing variation of speed. The function of a fly-wheel
is therefore to meet the fluctuations of demand for energy in
very short intervals of time, and it has no sensible effect in
regulating the variation of demand and supply over longer
intervals.

Gasholder Storage.—The distribution of gas is not
strictly a distribution of energy, but only of the means of
conveniently obtaining it. But a gas-lighting distribution
is analogous to a distribution of energy, and the demand
varies nearly as much as in an electrical distribution. The
gas engineer is happy in having a convenient and cheap
means of storage. Usually about twenty-four hours' supply
of gas is stored in the gasholders at a gas-generating station.

Hence the gas-making plant can be worked at an almost uniform rate day and night. Taking 25 c. ft. of gas as capable of yielding one effective h.p. hour of energy, it appears that gasholders cost about 5s. 6d. per effective h.p. hour stored. Mr. Trewby puts the cost of gasholders at a London station at 10,000l. per million cubic feet of gas supplied per day. Taking these gasholders to contain twenty-four hours' supply, and reckoning thirty cubic feet of gas per h.p. hour, the gasholders cost only about 6s. per h.p. hour of storage capacity.

A station supplying one million cubic feet per day, considered as a power station, works virtually at 1,666 average effective h.p. during the whole twenty-four hours. The cost of the gasholders adequate to meet any fluctuation of demand comes to only 6l. per average effective h.p. supplied. Allowing 10 per cent. for interest and depreciation, the storage adds about 12s. per effective h.p. to the annual cost of the power.

Hydraulic Storage.—Hydraulic storage will be discussed in another chapter. It is sufficient here to state that in hydraulic systems energy is stored in two ways, by hydraulic accumulators and by reservoirs. In the accumulators the total amount of energy stored is so small that it is only sufficient to meet momentary fluctuations of demand. The limitation is due to the great cost of accumulators, which amounts to something like 300l. per h.p. hour of storage capacity. The accumulator is like the fly-wheel of an engine. By pumping water to an elevated reservoir very large amounts of energy may be stored at not very great cost if local conditions are favourable. The pumped water descending again will re-store the energy by working hydraulic motors. A reservoir on an hydraulic system, like a gasholder on a gas supply, may be made large enough to completely meet all fluctuations of demand, so that the pumps can be worked at a uniform rate throughout the twenty-four hours.

Compressed Air Storage.—In systems for distributing energy by compressed air there is always more or less storage of energy, partly in special receivers, partly in the system of mains. The volume of compressed air, by expanding, gives up energy independently of a supply from the compressing plant. There is a fall of pressure, and there must be some limit, due to this fall, at which further energy cannot usefully be drawn from store, but must be supplied by the compressors.

For the purpose of diminishing fluctuations of pressure, small reservoirs of a capacity about equal to the air supply in three to five minutes are sufficient. For storage of energy much larger reservoirs are required. A very large air reservoir (400,000 c. ft. capacity) was at one time projected for the Paris compressed air system, but it was never carried out. It was intended that this should fill with water under a head of 260 ft. as the air was drawn off, the water being again driven out when the air supply from the compressors exceeded the demand. In that case the pressure would remain constant. More commonly the reservoirs merely supply part of the air they contain by expansion with diminishing pressure.

In the Portsmouth Dockyard compressed air system, for instance, there are eight air receivers of a total capacity of 18,000 c. ft. The normal (gauge) pressure is 60 lbs. per sq. in. Let $p_1 = 75$ lbs. (absolute); $p_a = 15$ lbs.; $v_1 = 18,000$. Then the work to fill the reservoirs, assuming isothermal compression and neglecting friction, would be—

$$144\, p_1\, v_1\, \log_e (p_1/p_a)$$
$$= 311,100,000 \text{ foot lbs.},$$

or 157 h.p. hours. This agrees sufficiently with a statement by Mr. Corner that the receivers can be filled by the 200 i.h.p. compressing plant in one hour. Not all this work can be recovered by calling on the store in the receivers. Suppose the pressure can be reduced to $p_2 = 40$ lbs. per sq. in. by gauge, or 55 lbs. absolute, before it is too low to work the motors. Then the work recovered would not exceed—

$$144\, v_1\, \{p_1\, \log_e (p_1/p_a) - p_2 \log_e (p_2/p_a)\}$$
$$= 125,600,000 \text{ foot lbs.},$$

or 63 h.p. hours. Part of this would of course be lost in inefficient action of motors. Mr. Corner states that about half the machines on the air mains can be worked for about two hours with air drawn from the receivers, the compressors being stopped. This means probably that they are driven at their ordinary intermittent rate of working for two hours. It is obvious that such an amount of storage as this, though it does not equalise supply and demand over the twenty-four hours, may have an important effect in regularising the working of the compressors, engines and boilers, and may not only be a convenience in permitting stoppage for slight repairs and in other ways, but may greatly reduce the waste of fuel and steam due to variation of demand for power.

Accumulator or Battery Storage.—The electrical engineer would be glad to have a means of storage equivalent to a gas-holder. For a time it was thought that such an equivalent had been found in the storage battery. The use of such batteries is limited to continuous current systems, and they have besides the practical defects—(1) that the maximum rate of discharge is limited, and (2) that about one-fifth of the energy stored is wasted. Nevertheless they would have been an extremely important factor in electric central station working but for their excessive cost. With a twenty-four hour load-line, such as that of most electric lighting stations, the amount of storage required to enable the generators to work at a uniform rate may be defined thus. The battery must be capable of supplying energy at a rate equal to three times the mean rate of supply for the twenty-four hours. Also it must be capable of storing during one part of the twenty-four hours, and re-storing in the other, about half the whole supply for the twenty-four hours. The cost of storage batteries prohibits their employment on this scale in large stations. Employed in a limited way, they serve some useful ends. In some stations they supply the energy required for ten to thirteen hours out of the twenty-four, during which time the engines are stopped. They diminish the fluctuation of load of the engines during the time in which they are running, storing energy not required in the external circuit. But they do not obviate the necessity for having a varying number of engines at work. Professor Kennedy puts the case well when he says that they 'enable the station to be shut down for some hours and act as fly-wheels, smoothing the irregularities of supply.' The accumulator battery, however, is inferior to the fly-wheel in the rate at which it will absorb and give out energy to meet momentary fluctuations of demand.

Cost of Accumulator Batteries.—From data given me by Professor Ayrton it appears that eight Epstein cells tested in the laboratory would work at one h.p. and store a charge for two and a half h.p. hours. The cells cost, without allowance for buildings, insulation or switching arrangements. or for waste of energy, 20*l.* That is, the bare cost of the cells amounts to 20*l.* per h.p. reckoned on their maximum rate of working, or to 8*l.* per h.p. hour stored. Suppose a station working at an average of 500 h.p. The maximum demand in the twenty-four hours would be 2,000 h.p., of which 1,500 would have to be supplied from the battery. The cost of the battery to supply energy at the necessary rate would be 30,000*l.* During twenty-four hours the quantity of energy supplied would be 12,000 h.p. hours, half of which must be stored. Batteries of sufficient capacity would cost 48,000*l.* Here the latter condition determines the cost. Taking interest at 5 per cent. and maintenance and depreciation at 12½ per cent., the annual cost of the battery would be 8,400*l.*, or nearly 17*l.* per h.p. of average rate of working of the station. This is the bare cost of the cells, without buildings, adjuncts or reserve.

In a project for lighting Frankfort-on-Main, Mr. Oskar von Miller and Mr. Lindley provided large secondary battery stations. The batteries had a capacity of 11,700 ampère hours, and were capable of supplying a current of 3,500 ampères at 100 volts. The batteries with wood platforms, insulation. &c., were taken to cost 25,100*l.*, and the buildings for them 11,600*l.* This is equivalent to a capital cost of 23*l.* per h.p. hour of storage capacity, or 78*l.* per h.p. reckoned on the assumed maximum rate of working. The difference between this and the previous calculations is that it includes necessary adjuncts, buildings, and reserve of storage to meet contingencies.

Thermal Storage.—Secondary batteries being too costly as a means of storage, except on a very limited scale, the question arises, Is any other means of storage available in conjunction with steam engines? Some means of hydraulic storage will be considered later : such means are rarely applicable for the storage of steam power. Lately Mr. Druitt Halpin has proposed a system of thermal storage which appears in many respects to meet the conditions required.

Energy is first obtained in steam power stations in the form of heat. Can the heat be directly stored? Heat is a very unprisonable form of energy, escaping through all bodies and in all directions. But in New York steam is transmitted through miles of pipes, and by reasonable jacketing the loss of heat is reduced to a moderate percentage of that carried. In a properly constructed storehouse for heat, with reservoirs closely packed and presenting little external surface, the radiation loss need not be large.

For storage, heat must be imparted to a material body of large heat capacity. It is easily given to water in boilers of ordinary construction. A body of water, highly heated in a well insulated chamber, will store a large quantity of heat. To permit the water to be heated it must be kept under the pressure corresponding to its temperature. Water heated above 212° and kept from vaporizing by pressure may be conveniently termed *superheated* water. The task of storing a mass of heated water presents no mechanical or physical difficulty.

It is a condition of any system of heat storage for central stations that the energy stored should be recoverable whenever and at any rate of supply required. Superheated water fulfils the condition. If the pressure is reduced, steam is generated instantly and in controllable amount. The steam generated can be used in the engines to produce mechanical energy as it is wanted.

Mr. Halpin's plan is therefore to communicate heat in boilers to a body of water. The heated water is stored in reservoirs under pressure. From the reservoirs steam is taken through a pressure-reducing valve exactly when and in what quantity it is required. Mr. Halpin proposes that the heat reservoirs should be under a pressure of 265 lbs. per sq. in. (absolute) when fully charged, the corresponding temperature being 406° F. He proposes that the steam engines should be worked at 130 lbs. per sq. in., corresponding to 347° F. The total heat stored when the reservoirs are fully charged is the difference of the total heat of the water at 406° and at 347° F., or the heat due to a range of temperature of 59°. Every pound of water falling in temperature through that range will yield 61 thermal units of heat. But the total heat required to generate a pound of steam at 130 lbs. per sq. in. from water at 347° is 868·8 Th. U. Consequently 14¼ lbs. of water falling in temperature from 407° to 347° will yield a pound of steam.

To allow for radiation loss and imperfect working, this may be taken at 16 lbs. of water per pound of steam. A simple cylindrical reservoir 8 feet in diameter and 30 feet long will contain 84,000 lbs. of heated water. Such a reservoir would be capable of generating under the conditions supposed 5,250 lbs. of steam at 130 lbs. per sq. in.

The steam consumption may be taken to be, per effective h.p., 18 lbs. per hour in condensing and 25 lbs. in non-condensing engines. Hence one such reservoir would store 286 effective h.p. hours if the steam is used in condensing engines, or 210 effective h.p. hours if the steam is used in non-condensing engines.

If the reservoir were fully charged and discharged daily it would yield 104,400 and 76,660 effective h.p. hours of stored energy yearly in the two cases.

A reservoir 30 ft. by 8 ft. would cost, erected with ample allowance for buildings and appendages, 470*l*. As it is not exposed to fire its deterioration would not be considerable, and 10 per cent. would be sufficient to cover interest, maintenance, and depreciation. Hence the first cost of such reservoirs reckoned on their storage capacity, and the annual cost per h.p. added to stored energy by the cost of storage, would be as follows:—

—	Cost of reservoirs per effective h.p. hour of storage capacity	Annual cost of storage per h.p. supplied from reservoirs
	£	£
Condensing plant .	1·64	3·94
Non-condensing plant	2·24	5·37

The cost in the last column is the cost due to storage of 8,760 effective h.p. hours annually.

Mr. Halpin's plans appear to be practicable and to promise considerable economy in stations where the load fluctuates greatly ; but they are untried, and it would not be fair to omit to point out that there are details of working involving difficulties which must be met. On the plan shown the steam must be generated in the tanks only, and a perfect circulation must be secured, perhaps by putting the storage tanks in series or groups. To prevent steam being generated where it is not wanted, and where it would be embarrassing, the temperature must not rise in any part of the system above the temperature due to the steam pressure in the steam space. Hence a large volume of circulation must be maintained. Other ways of working are however possible, if the method described proves to involve too much difficulty.

The cost on the mean annual h.p. supplied is not inconsiderable, but it is not prohibitive. The waste in irregularly working stations is so large that, *primâ facie*, it may be assumed that there is economy in storage on Mr. Halpin's system. But it must be remembered that this system attacks the boiler waste

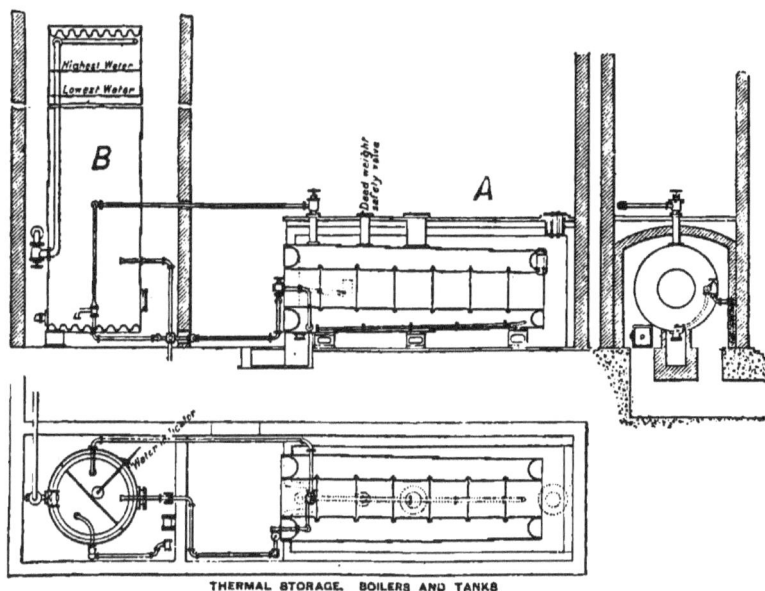

THERMAL STORAGE. BOILERS AND TANKS

FIG. 17.

only, and leaves the engine waste due to varying load untouched. To a certain extent the latter losses can be mitigated by subdivision of the engines.

Arrangement of Thermal Storage Reservoirs on Mr. Halpin's Plan.—It is possible that the best way of working thermal storage tanks is not yet known, but one arrangement proposed is shown in fig. 17. The steam boiler A is completely filled with water, the storage tank B nearly so. The two are in free communication by a system of circulating pipes. There is an ordinary feed pump supplying water direct to the boiler or the storage tank. But instead of keeping the water in the boiler at a nearly constant level, the level in the storage tank is kept nearly constant. In addition there is a circulating pump maintaining a rapid current of water from the boiler to the storage tank, and consequently back from the storage tank to the boiler. Water heated in the boiler is constantly being sent to the storage tank. and water cooled by disengagement of steam is returning to the boiler. The steam spaces of the tanks are all in communication. The pressure there will be the steam pressure due to the hottest tank. The steam required is taken off through a reducing valve. It will then be generally dry or slightly superheated, in consequence of wiredrawing, which is advantageous for the efficiency of the engines.

In a station with thermal storage tanks, the boilers would be of a size sufficient to supply the mean demand for steam on the day during the year when the demand is greatest. The boilers would be worked continuously at a nearly uniform rate, like a bank of gas retorts. The heat not required in hours of small demand would be stored in the tanks. The excess of heat required to generate steam in hours of great demand would be taken from the store in the tanks.

Comparison of Steam Central Stations worked in the ordinary way and worked with Thermal Storage.—The following example of a central station, with a load-line like that of the Kensington Electric Lighting Station, has been worked out by Mr. Halpin. Fig. 18 shows the ordinary midwinter load-line of such a station, the ordinates representing effective h.p. The total output in twenty-four hours is 15,600 h.p. hours, so that the mean rate of working is 15,600/24, or 650 effective h.p. For practically seven hours the demand exceeds the mean demand, and during that period altogether 10,450 h.p. hours must be supplied, which is 5,900 h.p. hours in excess of the $7 \times 650 = 4,550$ h.p. hours which correspond to the average demand in the twenty-four hours. The maximum demand for a short period is at the rate of 2,400 h.p., or 3·7 times the mean rate.

If this station is worked in the ordinary way, without any storage of energy, it must have engines and boilers able to develop 2,400 h.p. These will be worked to their full capacity for a short time only, and during seventeen hours they will have to develop less than 650 h.p.

In a station with adequate thermal storage tanks, the engines would have to be of 2,400 h.p., but boilers of 650 h.p. only would be necessary. The thermal storage tanks would supply during seven hours the excess energy, amounting to 5,900 h.p. hours, which would have been carried into them during the seventeen hours of small demand. Taking 16 lbs. of water to supply 1 lb. of steam, 360 lbs. of water stored would supply one effective h.p. hour. Then the thermal storage tanks would have to contain 900 tons of superheated

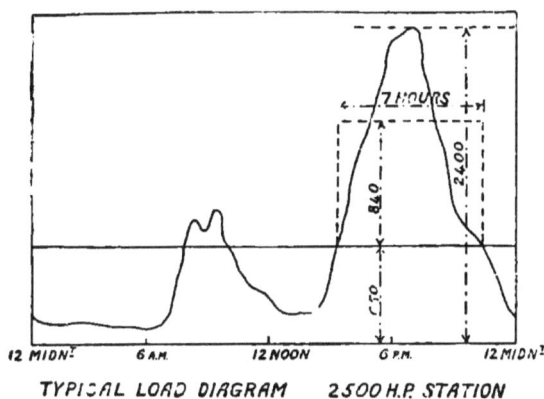

TYPICAL LOAD DIAGRAM   2500 H.P. STATION
Fig. 18.

water. Twenty-four tanks 30 ft. × 8 ft. would have sufficient capacity.

Station without storage tanks.—Eight boilers, each with 2,000 sq. ft. of heating surface, and two boilers in reserve. Cost of ten boilers with pipes, pumps and cost of erection, 10,300*l*.

Station with thermal storage tanks.—Two similar boilers, and one in reserve. In addition, 24 storage tanks. Cost of three boilers and 24 tanks with pipes, pumps and cost of erection, 15,000*l*.

The cost of the plant is greater with the storage tanks by 4,700*l*. Taking interest and depreciation at 10 per cent., this would correspond to an annual charge of 470*l*. The extra cost of the storage is sixteen shillings per h.p. hour of storage capacity, or about nine shillings per annum per h.p. of average output from the station. The extra annual charge of 470*l.* is about the value of 500 tons of coal at London price. From data given above it will be seen that at existing electric lighting stations the consumption of coal is at least 4 lbs. per effective h.p. hour. This for the station under consideration would correspond to an annual consumption of 17,000 tons of coal. If the saving of waste in the thermal storage station due to working the boilers regularly instead of irregularly amounted to only one half-pound per effective h.p. hour, the annual saving of coal would be 1,270 tons.

Cases still more favourable for the application of the thermal storage system are those where heat is now absolutely thrown away. The destructor for ash-bin refuse, which has already been described, must be worked continuously day and night. This makes it difficult to utilise the heat. But with thermal storage tanks the heat might be captured and stored for use at hours when mechanical work had to be done. In such a case the advantage of thermal storage would seem to be very great. One other similar case has been thought of by Mr. Halpin. In the production of lighting gas about 312 lbs. of coke are burned per ton of coal carbonised, and about 6,240 lbs. of furnace gases escape from the retort bank at a temperature of 1,200° F. per ton of coal carbonised. The heat in these products of combustion is now entirely thrown away. If they were taken through the flues of a boiler, they might be reduced to 600° before escaping. They would furnish about 759 lbs. of steam per ton of coal carbonised, or about 25 effective h.p. hours.

The gas industry is a very extensive one, and the aggregate waste of heat is enormous. At present the chief difficulty in utilising it is that there is no continuous work requiring to be done. But if the heat can be stored and used when wanted the case is very different.

Since these lectures were delivered Mr. Halpin has pointed out some modifications of his system. The plan described above of storing superheated water is a system of complete storage permitting the boilers to be worked continuously at average load. But by merely heating a quantity of feed water sufficient to supply the boilers during the period in which the load exceeds the mean load, an amelioration of the conditions of working is secured. This may be termed partial thermal storage. The feed is heated to the temperature of the boiler steam only, and the difficulties which may attend the use of very high pressures in the storage tanks are avoided. A subordinate advantage is that the feed water in the storage tanks may be made to deposit part of its lime salts in the tanks, where it is less injurious than in the boiler, and more easily removed.

CHAPTER V

WATER POWER

WHERE there exists a natural waterfall with a considerable and regular flow, and where natural conditions are favourable for the construction of the necessary works, water power is generally much cheaper than steam power. The water costs nothing; the cost of maintenance of the hydraulic machinery and the cost of superintendence are small. The annual cost of power consists almost entirely of interest charges on the capital expended in works and machinery. The power obtained in this way is regular, controllable and convenient.

So well is this recognised that industries like those for the electric reduction of metals, which involve the expenditure of large quantities of mechanical energy, seek localities where water power is available as a first condition of successful operation. In this country there is comparatively little water power. The drainage areas are comparatively small, and the flow of the streams is irregular. Coal also is cheap and abundant. Hence water power is of secondary importance. But in some other countries water power is hardly subordinate to steam power in manufacturing industries. The extension of methods of transmitting power to a distance would make many natural waterfalls available which are at present unutilised. It is because the transmission of power is now receiving so much attention that there is a remarkable revival of interest in the utilisation of water power.

With few exceptions, water power has hitherto been employed only in the immediate neighbourhood of the natural waterfall. Where there has been a distribution of water power to different factories, it has commonly been effected by conveying the water itself to the factories, where the power is developed and used. In some cases water has been conveyed for mining and manufacturing purposes very considerable distances. But a more convenient and cheaper method of transmitting power derived from water would greatly increase the availability of this source of energy, and, in some countries, would change the relative importance of steam power and water power.

It appears from a report by Mr. Weissenbach that, in 1876, 70,000 h.p., derived from waterfalls, were in use for manufacturing purposes in Switzerland. It is estimated that the total available water power in Switzerland amounts to 582,000 h.p. Putting the annual value of a h.p. at 6l., this corresponds to a total annual value of 3½ million pounds. Supposing it used to replace steam power, there would be an annual saving of 1¼ million tons of coal. It is stated that, at the present time, Switzerland pays annually to other countries 800,000l. for coal.[1] The greatest part of this could be saved, if its natural wealth of water power were rendered available. The recognition of the importance of water power is now exciting great interest in Switzerland, and many factories are either using water power, or making preparations to do so.

The utilisation of water power often involves the construction of large permanent works, such as river dams, reservoirs, and canals. Mr. Emery estimates that at Lawrence, in the United States, 200,000l. was spent on works, independent of the hydraulic machinery, and at Lowell a still larger sum.[2]

Such extensive works can be best executed by an association, in the interest of many consumers. Thus is created a water-power company, who establish what is virtually a Central Water Power Station, and a distribution of power at a rental to consumers. In the American cases, as already stated, the water itself is distributed in canals to consumers, at a level permitting the creation of a waterfall at the mill or factory. But in certain other cases a further step is taken: the Water Power Company utilise a natural fall, and erect the necessary turbines, and then transmit the power in the form of mechanical energy to consumers. Installations of such a kind, now of a quite respectable antiquity, were erected at Schaffhausen, Freiberg,

[1] Reifer. *Berechnung der Turbinen.*

[2] 'Cost of Steam Power,' C. E. Emery; *Trans. Am. Soc. of Electrical Engineers*, vol. x. p. 123.

Zürich, and Bellegarde. In these cases the method of transmitting power adopted, admirable as it was, had limitations, and the extension of the works was restricted. Now that there are new means of transmission, the Schaffhausen and Zürich power-generating stations are being increased, and a new and remarkable installation has been erected at Geneva.

The original project for utilising the motive power of the Rhône at Geneva, partly for pumping a supply of water, partly for motive power for industry, comprised 20 turbines of 300 h.p. each, or an aggregate of 6,000 h.p. Fourteen of these were at work in 1892. Four more of somewhat larger size will, it is expected, be constructed by 1898. When these are at work the whole available water power in Geneva will be utilised. But it is foreseen that the demand for power will not then have been satisfied. The total receipts for the installation reached 22,500l. in 1891, and were increasing 2,200l. annually. The municipality of Geneva has determined to provide for future demands, and plans are being studied for utilising 12,000 h.p. at a point on the Rhône six kilometres below Geneva, whence the power will be distributed electrically. At Biberist, near Soleure, Messrs. Cuenod, Sautter, & Co. have utilised 360 h.p., and transmitted it 28 kilometres electrically. At Genoa, water power, due to surplus fall along a line of water main, has been utilised at three stations. The greater part of the energy is transmitted to Genoa for electric lighting and power purposes.

These are cases where water power has been utilised and distributed which are actually in operation. But many other schemes, some of them on a still larger scale, have recently been projected. On the United States side of the Falls at Niagara an immense work is in progress for obtaining 100,000 h.p., and distributing it electrically. The rock tunnel for this amount of power is completed. A turbine wheel pit for three 5,000 h.p. turbines is nearly complete, and another for turbines of 6,000 h.p. intended to drive a paper mill. A project for utilising a still larger amount of power on the Canadian side is under consideration. There has been a project for utilising 10,000 h.p. on the Dranse at Martigny ; another for utilising 20,000 h.p. at a point 17 kilometres above Lyons. In Sweden there is a project to transmit power from the Dal River at Mansbo to the Norberg mining district, a distance of 10 miles. There is a pro-

ject to transmit power from a fall on the Judal River to Otter-
sund, a distance of 11 miles. There are projects to transmit
power, from the falls of Trollhatten and the river Motala, to
Gothenburg and Nordköping. It is stated that works are
actually in progress for obtaining 11,000 h.p. at Orizaba, in
Mexico, on a fall of 115 feet. The power is to be distributed
electrically to factories.

Water Power in the United States of America.—It is in the
United States of America that water power is most largely used,
where it is in most direct competition with steam power, and
where data for a comparison of their relative advantages can
best be obtained. Interesting information as to the extent to
which water power is utilised in the United States is given in a
paper by Mr. G. F. Swain read before the American Statistical
Association.[1]

The money value of the water power utilised in the United
States is very considerable. From the returns of the Tenth
Census it appears that, in 1880, there were 55,000 water-wheels
and turbines, of an aggregate of 1,250,000 h.p. At 5*l.* per h.p.
per annum the water power utilised is worth 6,250,000*l.* a
year.

The comparison of the relative amount of water and steam
power is interesting. Taking the whole of the United States,
36 per cent. of the power used in manufacturing was, in
1880, water power, and 64 per cent. steam power. In certain
industries the proportion of water power was greater. In the
manufacturing of cotton and woollen goods, of paper and of
flour, 760,000 h.p. derived from water and 515,000 h.p. derived
from steam were employed. In the North Atlantic division
4·81 water h.p. are utilised per square mile.

Division	Water power per cent.	Steam power per cent.
N. Atlantic	43·1	56·9
S. Atlantic	49·5	50·5
N. Central	22·2	77·8
S. Central	22·5	77·5
Western	35·3	64·7
The United States	35·9	64·1

[1] 'Statistics of Water Power employed in Manufacturing in the United
States,' by G. F. Swain; *American Statistical Association*, March 1888.

Fig. 19 is a map, taken from Mr. Swain's paper, which shows that over a considerable area of the United States the water power used exceeds the steam power. It should, however, be pointed out that, in the decade 1870–80, during which the total power used increased 45 per cent., 9 per cent. of the increase was due to water power and 91 per cent. to steam power. It is

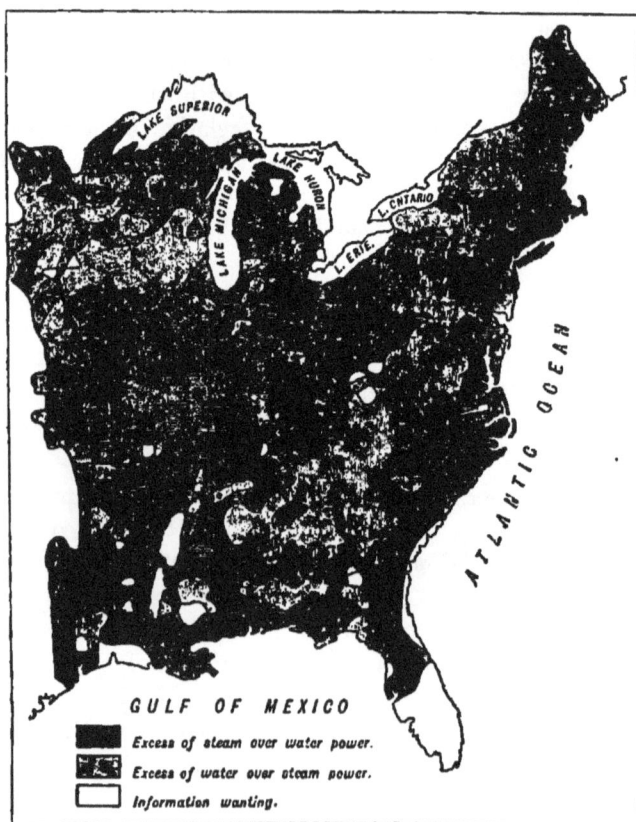

FIG. 19.

possible that, under the new conditions now obtaining, the present decade will show a greater relative increase of water power.

American Method of Distributing Water Power.—The method in which water power is distributed in America to a number of consumers is almost peculiar to that country. A Water Power Company is formed which undertakes the construction of the permanent works, such as a river dam, sluices, and distributing canals. In New England, there are five water power stations where more than 10,000 h.p. is utilised during working hours, and thirteen stations where more than 2,000 h.p. is utilised. The water is distributed to mill-owners, who construct the turbines and pay a rental to the Water Power Company proportioned to the amount of water used. The earliest application of this system was at Paterson, New Jersey, where the Passaic River furnishes about 1,100 h.p. night and day.[1] At Lowell, Massachusetts, the utilisation of the water power began in 1822. The Merrimac River has a fall of 35 feet, and furnishes at the minimum about 10,000 h.p. during the usual working hours. At Cohoes, in the State of New York, the Mohawk River has a fall of 105 feet. It could furnish about 14,000 h.p. during working hours, but is only partly utilised at present. At Manchester, New Hampshire, the Merrimac has a fall of 52 feet and furnishes at the minimum about 10,000 h.p. during working hours. At Lawrence, Massachusetts, the Essex Company built a dam, forming a fall of 28 feet, and obtaining a minimum power of 10,000 h.p. during working hours. At Holyoke, the Hadley Falls Company built a dam, forming a fall of 60 feet and rendering a power of 17,000 h.p. available during working hours.

To indicate the magnitude of some of these works it may be stated that at Lawrence the masonry river-dam is 900 feet long and 32 feet in height. The cost was 50,000*l*. From this dam two canals extend down stream, one on each bank, and between these canals and the river are located the mills, occupying the entire river front. On the north side the mills extend for a distance of more than a mile. The cost of the canal on the north side, 5,330 feet in length and 100 feet in width at the upper end, was 50,000*l*. The canal on the south side, 2,000 feet in length and 60 feet in width, cost 30,000*l*.

The case of the town of Holyoke, Massachusetts, may be described in somewhat greater detail. All the factories in the town are worked by water power, and the system is strictly a distribution of power from a common source to many consumers, at a rental proportional to the power used, although the power is actually developed in the mills by turbines belonging to the mill-owners. The Holyoke Water Power Company controls the

---

[1] J. B. Francis ; *Trans. Am. Soc. of C. E.*, vol. x. p. 189.

flow of the Connecticut River. which has a drainage area above
the town of 8,144 square miles. The first weir or dam was
built in 1847, but it was carried away. A second dam of crib-
work was built in 1849. In 1868, an apron was constructed to
protect the rock immediately below the dam. Since then, Mr.
Clemens Herschel has carried out extensive repairs of the dam [1]
under conditions of singular difficulty with great success. The
structure is now 130 feet in width, 30 feet in height, and 1,019
feet in length. From above the weir, a canal supplies water to
the highest line of mills. After driving turbines in these mills,
the water is collected in a second canal. which is a supply canal
to a second line of mills. The tail-races of these mills discharge
into a third canal, from which other mills are supplied before
the water returns to the river at a point where the level is 60
feet below the level above the dam.

Nearly 15,000 h.p. are in use by day and over 8,000 h.p.
at night, of which part are 'permanent powers' held under
leases and subject to annual rental; the balance are 'surplus
powers' held by contracts subject to withdrawal at short notice.
There are 139 turbine water-wheels in Holyoke, of which 59
run about ten hours daily, and 80 run from Sunday midnight
to Saturday midnight, or 144 hours per week.

With the grant of land for a mill there was leased the right
to use a definite portion of the water power. A 'mill-power'
is defined as 38 cubic feet per second, on a fall of 20 feet, during
sixteen hours per day. This is equivalent to about 63 effective
h.p. on the turbine shaft. At the time when Mr. Herschel
became engineer to the Water Company, the water was used
extravagantly. By introducing a system of testing the turbines
before their erection at the mill, data were obtained from
which the quantity of water used by the turbine, at any gate-
opening for any head, could be calculated. At each mill, gauges
are placed showing at any time the height of the fall. Observations of the fall and gate-opening at each turbine are made by
inspectors daily, and the quantity of water used is thus ascertained. The excess of water used above that granted by the
mill lease is charged for as ' surplus power.'

It is an advantage to the mill-owners to have this surplus

[1] ' On the Work done for the Preservation of the Holyoke Dam in 1885,' by
Clemens Herschel ; *Trans. Am. Soc. of C. E.*, vol. xv. p. 543.

power at moderate cost, and the system introduced by Mr.
Herschel, by which the charge for water is made to depend on
accurate measurement of the amount used, led to great economy
in the use of the water and secured a large surplus power at
the disposal of the mills.

Observations of the gate-opening of each turbine and the
fall between head and tail race are made at each mill, at least
once in the day and once at night. Three inspectors are engaged exclusively in this work. From the daily observations
the quantity of water used at each mill is calculated. Part of
this is charged for, according to the terms of the lease, at a fixed
rental. The balance is charged for as surplus power. In times
of drought the amount of surplus power is restricted.

The Holyoke Testing Flume.[1]—The first flume at Holyoke
was constructed by Mr. Emerson. The business of testing
turbines became so important that the Power Company erected a
new flume and buildings, under the direction of Mr. Herschel, in
1882. This is placed between two of the canals of the Company
where from 17 to 19 feet of head is available. It has offices
and repairing shops, and is fitted with dynamometers of different
sizes, a measuring weir with accurate hook gauges, gauges for
measuring the head, clocks with electric signal bells and other
appliances for experiment.

Fig. 20 shows the general arrangement of the flume. The
water enters through a 9-foot wrought-iron pipe A from the
upper canal into a masonry ante-chamber, from which it is
admitted to the chamber c by the regulating sluices GG. In
this chamber there is a small turbine for working the repairing
shops. I is a tail-race for this turbine. From c the water
passes over stop planks into the wheel chamber D, on the floor
of which is placed the turbine to be tested. For cased wheels,
the chamber D is empty of water, and a supply pipe is attached
to the timber bulkhead L. The water is discharged through
culverts N into the tail-race E, at the end of which is the measuring weir o; R is a suspended platform over the tail-race. At P
is a recess in which the measurements of the water level over
the weir are made.

The turbine to be tested being fixed in the chamber D, a

[1] See a paper by R. H. Thurston, *American Inst. of Mechanical Engineers,*
1887.

friction brake with water-cooled rim is fixed on its shaft. Attendants at the brake adjust the weights and the tension of the friction band. Other observers at the weir note the height of water over the weir. The revolutions of a counter are noted at minute intervals. A series of trials are made with each gate-opening, the load on the brake being varied to give different speeds. From the observations there can be plotted curves giving the discharge for each gate-opening at different speeds, and the efficiency corresponding to each gate-opening and speed. It is not absolutely necessary that the turbine should be tested on the fall on which it is to be used, because the discharge very approximately varies as the square root of the head, and the efficiency does not vary much for different heads when the speed is proportional to the square root of the head.

The measuring weir has a capacity of 200 c. ft. per sec. Any head from 4 to 17 feet can be used. The cost of a test is 10 per cent. on the list price of the wheel, with a minimum charge of $30. In 1883 the number of wheels tested was 185.

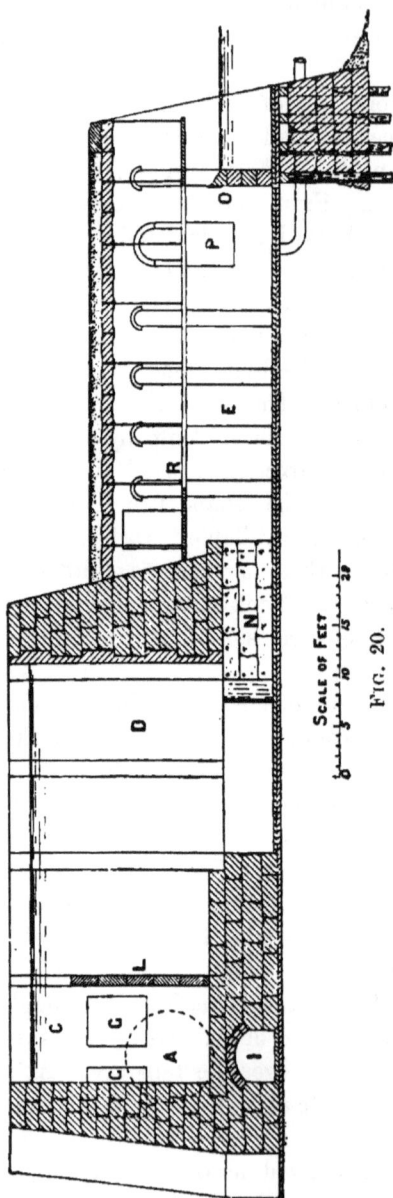

Relative Cost of Water and Steam Power in the United States.—
In some cases local conditions are so favourable that water power
can be developed at an almost nominal cost. In other cases,
with less favourable local conditions or from unforeseen contingencies, expenditures have been incurred which have made the
cost of water power excessive, greater in fact than that of steam
power.

Mr. Swain puts the average cost of steam power in the States
in favourable localities at 4*l.* per h.p. per annum, and that
of water power at about 2*l.* per h.p. per annum. Both these
estimates are so low that it may be suspected that they are
based rather on the nominal power of the plants than on the
average actual h.p. used throughout the year. The cost of water
power however varies greatly. Mr. Swain states that, while in
the North West of the United States the cost for interest, depreciation and water rental is about 2*l.* 2*s.* to 2*l.* 5*s.* per h.p. per
annum, in New Jersey it is from 12*l.* to 15*l.* That water power
is used at all at a cost so large as this proves that it has advantages of convenience, compared with steam power, which balance
some excess of cost.

It is somewhat difficult to arrive at a precise knowledge of
the cost of water power in the great works in America, because
of the gradual way in which they have been developed and the
want of complete data as to the amount expended. Mr. C. E.
Emery,[1] who is probably rather prepossessed in favour of steam
power, has made the following estimate of the cost of water
power at Lawrence.

He puts the total cost of the structural works at Lawrence
at 200,000*l.*, and the power utilised as equivalent to 13,000
h.p. for ten hours daily throughout the year. That makes
the cost of structural works 15·4*l.* per h.p. The cost incurred
by the mill-owners in erecting turbines, sluices, &c., he puts
at 9*l.* per h.p. of the turbines, or 13*l.* per average h.p. actually
utilised, the turbines being generally constructed to yield
surplus power in times of emergency. This makes the total
expenditure 28·4*l.* per average h.p. utilised ten hours daily
throughout the year. He allows 2½ per cent. for depreciation,
1¼ per cent. for repairs, 1¼ per cent. for taxes, 10 per cent. for

[1] 'The Cost of Steam Power,' by C. E. Emery: *Trans. Am. Soc. of Electrical
Engineers*, March 1893.

interest, and 2 per cent. for working expenses, or altogether 17 per cent. on the capital expenditure. This makes the annual cost of a h.p. at Lawrence 4·7l. per annum, which he takes to be about the same cost as that of steam power, with economical engines, and coal at 8s. to 12s. a ton.

This estimate is for cases where only a moderate fall is available ; with large falls and in favourable conditions water power can be obtained at a much less cost.

Cost of Water Power at Geneva.—It appears that at Geneva for the first groups of turbines erected, of 840 h.p., and for the river works then completed, the capital cost amounted to 60l. per effective h.p. The groups of turbines subsequently erected have cost only 19l. per h.p. The mean cost, when the present works are completed, will amount to 27l. per effective h.p. In this case the water costs nothing. If we allow 5 per cent. for depreciation, repairs, and working expenses. and 10 per cent. for interest on capital, the cost per h.p. per annum will only amount to 4l. In the new works, below Geneva, where 12,000 h.p. are to be utilised, it is estimated that the whole cost for turbines and structural works will amount to 60l. per h.p. for the first 2,400 h.p. When the whole installation is completed, the capital cost will be only 27l. per h.p.

STORAGE OF WATER POWER

The need of storing power obtainable from a natural water-fall arises out of considerations which are different from those which apply in the case of steam power. A river flows day and night with an energy which varies seasonally, but not from hour to hour. The work to be done in a factory or central station varies necessarily from hour to hour, and in the majority of cases there is no demand for mechanical energy for half the twenty-four hours. If no means are provided for storing the available energy, a large part flows away, and is wasted.

There is another reason. In the case of water power nearly the whole cost is due to interest on capital expended on permanent structures, and an allowance for depreciation. Very little is due to working expenses. But with steam power only about one-third to one-half of the whole cost is due to permanent charges, and two-thirds to one-half is due to wages and fuel. If a steam engine stops for twelve hours out of the twenty-four, half the coal and wages are saved. Though the cost per h.p. hour is increased. it is only increased by about 25 per cent. But if water-power machinery is stopped for half of the twenty-four hours, the cost of a h.p. hour is doubled.

At some of the American water-power stations, an inducement is held out to consumers of power to work night and day, a lower rate being charged for power taken at night. In other cases the difficulty is met by storing the water during the night so that it can be used during the day. The amount of power available in working hours is then doubled. One of the most characteristic advantages of water power is that the storage of energy is possible by means not excessively costly or difficult. Further, it is the facility of storing energy in elevated reservoirs which, in some cases, makes it profitable to pump water to be afterwards used for power purposes.

There are two distinct methods of storing energy in hydraulic systems—accumulator storage, and reservoir storage.

Perhaps, on superficial consideration, it would appear very unlikely that it could be profitable to pump water for power purposes by steam pumps. There are cases where it is so. One of these is the system of hydraulic high-pressure transmission devised by Lord Armstrong. This system is used, and can only be used advantageously, to work a great number of intermittently working machines. A single steam engine, working almost continuously, pumps water which actuates a great number of intermittently working hydraulic motors. Naturally the fluctuation of demand for power varies a great deal, and storage is almost essentially necessary. Perhaps it is to the invention of the accumulator, a means of storing the energy of pressure water, that the success of the system of hydraulic transmission is chiefly due.

The hydraulic accumulator is simply a vertical cylinder, with a heavily loaded plunger, into which the water is pumped till it is required, and from which it is discharged by the descent of the plunger.

Let A be the area of the plunger in sq. ft.; P the total load on it in lbs. Then $p = \text{P}/\text{A}$ is the pressure at which the water is delivered in lbs. per sq. ft. If h is the length of stroke of the accumulator plunger, in ft., then Ah is the greatest quantity of pressure water it will store, and

$$p \text{ A} h \text{ foot lbs.}$$

is the energy stored when it is fully charged.

The pressure used in systems of hydraulic transmission is generally 750 lbs. per sq. in. Now one of the very large accumulators of the London Hydraulic Power Company has the following dimensions: diameter of plunger, 20 ins.; stroke, 23 feet; at 750 lbs. per sq. in., this accumulator, large as it is, stores only 2·4 h.p. hours, a comparatively insignificant quantity. The cost of this accumulator, reckoned on the capacity for storing energy, must be very large indeed. What makes the accumulator so important is that its rate of discharge is very great. It would probably supply 100 h.p. for 1½ minutes. Hence, like the fly-wheel, the use of the accumulator is limited by its costliness to meeting fluctuations of demand for energy in short periods of time. It cannot be used to average the demand and supply during long periods. This must be effected by varying the engine power which supplies the energy. Costly, like the electric secondary battery, it has an advantage over the latter, that the rate of discharge is unlimited.

If a suitable elevated site can be found, then reservoirs can be built of very large capacity at a cost not large per cubic unit stored. Let A be the mean surface area of the reservoir, h the variation of depth of water in the reservoir, and H the mean height of the reservoir water level above the hydraulic motors supplied. Then the volume of water in the reservoir when full is Ah cubic feet. The gross amount of energy stored, not allowing for loss in pipes and motors, is

$$\text{G A H } h \text{ foot lbs.}$$

(G $= 62 \cdot 4$, the weight of a cubic foot of water), or—

$$\text{A H } h / 31,740 \text{ h.p. hours.}$$

At Zürich, for instance, the storage reservoir contains 353,000 c. ft., at an elevation of 475 feet above the motors. It therefore stores 5,284 h.p. hours.

At both Geneva and Zürich very remarkable and extensive systems for utilising and distributing water power have been for some time in successful operation. In both cases the water flowing out of a large lake with a comparatively small fall is utilised to furnish a very considerable and valuable power. In both cases it has been found convenient and economical to use the low pressure turbines, in the rivers flowing out of the lakes, to pump water to a reservoir at a great elevation, and to use the pumped water for ordinary motor purposes.

Data may be taken from the Geneva installation, though the works in both cases are similar, and both have been financially successful.

At Geneva, the Rhône, flowing out of Lake Leman, has by skilful arrangements been made to afford a clear fall varying from $5\frac{1}{2}$ to 12 feet. There have been erected in the river, on this fall, or shortly will be erected, 18 low pressure turbines, giving in the aggregate over 6,000 h.p. These turbines are used primarily in pumping a supply of potable water for Geneva. But, since 1871, there has grown up in Geneva a system of using water from the town mains for motor purposes, and it is an important secondary function of the low pressure turbines in the river to pump water used for motor purposes. To begin with, it may be pointed out that the low pressure turbines, together with the building and permanent works required to create the fall, are very costly. It is desirable that, to reduce the permanent charges on the energy furnished, they should work as long hours as possible. Further, the total power furnished is even now insufficient for the work to be done, and it is necessary that water power should not flow to waste during the night hours. But, even for other reasons, it became necessary to use storage of energy, before the present exigencies arose. In the earlier period of the enterprise, to maintain constant pressure in spite of the fluctuation of demand on the mains, it was necessary that the turbines should be constantly pumping in excess of the demand. The surplus was discharged through a relief valve, and involved a constant waste. To meet fluctuations of pressure, four large air vessels were constructed 5 ft. in diameter and 39 ft. high. With these, in conjunction with the relief valves, the working was smooth and successful, but necessarily wasteful.

When the electric installation was erected at Geneva and driven by turbines actuated by the pumped water, the necessity of storage become more evident. At 4 kilometres from Geneva a site was found at an elevation of 390 feet above the lake, where

a reservoir could be constructed. The reservoir contains
453,000 cubic feet, and stores therefore 5,573 gross h.p. hours
of energy, which can be accumulated and discharged daily.
Allowing for the loss at the motors driven by the water,
the effective energy stored may be taken at 4,180 h.p. hours.
The reservoir is a covered reservoir of expensive construction;
but its cost does not exceed 2·4l. per h.p. hour of energy
stored. It is a work requiring little maintenance, and it hardly
adds 3 shillings per annum to the cost of a h.p. supplied. On
the other hand the energy stored would without it have gone to
waste, and for this a rental is obtained of 8l. per annum.

There is now in London an admirable system of hydraulic
power distribution, perfectly adapted to the special purposes
to which it is applied and mechanically and financially successful. But it is a system of limited applicability. Large as it
is, the number of renters of power is under 2,000, and the maximum pumping capacity of the supply stations at present erected
is 2,600 effective h.p.

In the comparatively small town of Geneva, with one
eightieth of the population of London, there is a system of power
distribution as large as, and of more varied applicability than,
the system in London. There were in Geneva three years ago
137 hydraulic motors aggregating 280 effective h.p. on the low
pressure mains, and 79 motors aggregating 1,284 h.p. on the
high pressure mains. The use of the hydraulic power was increasing at the rate of nearly 200 h.p. per annum. Lastly, the
power in Geneva is distributed to ordinary consumers at one-
fifth, and for the electric lighting station at one-twelfth, the
London price.

Perhaps it is fair to add that in London there are 2,500 gas
engines, which represent a considerable aggregate power virtually supplied from a central station. Nevertheless motive
power is more generally and more cheaply distributed in Geneva
than in London. No doubt local conditions have favoured the
adoption of plans for distributing power in Geneva, but perhaps
it has not yet been fully recognised in London what an advantage cheaply distributed power is. When this is better
recognised means may be found to make motive power more
available as a purchasable commodity.

CHAPTER VI

HYDRAULIC MOTORS

A MOVING stream of water has, in the most general case, velocity, pressure, and elevation, relatively to some level to which it is descending. Its total energy per pound is the sum of its velocity head, its pressure head, and its elevation head. If v is the velocity in feet per second, p the pressure in pounds per sq. ft., h the elevation above the level to which it can descend, and G the weight per cubic foot, the total head is—

$$H = \frac{v^2}{2g} + \frac{p}{G} + h \text{ feet}$$

which is numerically equal to the total energy of the fluid per pound. Hydraulic motors are generally arranged to utilise one or other of these components of the total energy.

(*a*) The elevation head may be utilised by allowing the water to descend from its highest to its lowest level in the buckets of a water-wheel. The wheel is driven by the weight of the water on the loaded side of the wheel. Gravity water-wheels are motors of an antiquated type, which have been almost completely superseded by others either more efficient or less costly.

(*b*) The pressure head may be utilised by allowing the water to descend in a closed pipe at a velocity of 1 to 3 feet per second and then to drive a piston. The pressure on the piston is that due to the height of the column of water between the highest and lowest level, less some frictional losses which are comparatively insignificant.

(*c*) A third way of utilising the energy of water is to allow the water to issue in a jet with the kinetic energy due to the whole height of the fall. The jet is deviated by the curved

floats of a turbine wheel. The effort driving the wheel is numerically equal to the change of momentum per second in the jet,
reckoned in the direction of motion.

Hydraulic Pressure Engines.—Pressure engines with pistons
driven by water are necessarily slow-moving, and consequently
are cumbrous and expensive machines, except in the case where
very great pressures have to be utilised. In two classes of conditions motors of this kind are used with advantage : (1) For
working cranes, lifts, &c., where slow and steady motion is necessary, direct-acting water-pressure cylinders are safe and convenient; they are so convenient that it is even advantageous to
obtain pressure water by pumping, in order to work cranes and
lifts in this way. (2) Small rotative pressure engines are often
used where water under moderately high pressure is available,
especially for driving machines used intermittently. Their convenience in such cases counterbalances the defect of a low average
efficiency, and they serve also as almost perfect meters of the
amount of water used.

The volume of water used in a pressure engine cylinder per
stroke is constant, whether a small or a great effort has to be
exerted. The engine uses as much water for light loads as for
heavy loads. Motors of this kind have often an efficiency of 80
per cent. at full load. But with half load the efficiency would
be less than 40 per cent. Hence with a varying load they are uneconomical. Various devices have been tried for obviating this
defect. For cranes, three pressure cylinders are sometimes provided, which can be used one, two, or all three together, according
to the weight to be lifted. Lord Armstrong invented an
arrangement of this kind many years ago, and a similar arrangement is used in some London warehouses. A three-power 20
cwt. crane may have cylinders suitable to raise 4, 12 or 20 cwts.,
according as one, two, or all three cylinders are in action.

In Switzerland, small hydraulic rotative pressure engines,
made by Schmid of Zürich, have been extensively used in systems
of hydraulic distribution of power. Fig. 21 shows one of these
engines. Such engines have one or two oscillating cylinders.
the movement of which opens and closes the cylinder ports, so
that no separate valve gear is required. They use the same
quantity of water per revolution whether the load is light or
heavy. These motors are cheap and very convenient for small

powers. A revolution counter is attached to each motor, and the charge for water is based on the counter readings. The friction of the engine with 100 feet of water pressure is only 5 per cent., and the efficiency at full load is 80 per cent., apart from frictional losses in the supply pipe.

The revolving engine of Mr. Arthur Rigg, which is sometimes used as a steam engine, can be equally well used as a water-pressure engine. This can be provided with a very ingenious arrangement for varying the stroke, so that the quantity of water used can be adjusted to the amount of work to be done. The engine may have, therefore, a nearly equal efficiency at full

Fig. 21.

and light loads. The clearance varies with the variation of stroke, but in the case of an hydraulic engine this does not affect the efficiency. From its nearly perfect balance, it can be run at high speeds.

Fig. 22 shows a four-cylinder Rigg Hydraulic Engine, which has been working a tramway in North Wales for two years, driven by a natural head of water, which gives a pressure of 206 lbs. per sq. in. at the engine. The gunmetal cylinders are each $6\frac{1}{2}$ ins. diameter, and 9 ins. stroke. It works at 90 revolutions per minute with full load, at 110 revolutions with half load, and at 130 revolutions drawing an empty waggon.

The four cylinders are pivoted on a central stud. Their plungers act on four crank pins, on a revolving crank-disc. The length of stroke of each plunger is twice the eccentricity of the central stud. By varying this eccentricity by means of the subsidiary hydraulic cylinders, shown in fig. 22c, the stroke of the plungers is varied to suit the load. A port to each cylinder is provided in the central boss, by which it is pivoted on the central stud. This boss revolves against a valve face having an inlet port on one side, and an exhaust port on the other. The cylinders receive pressure water during one half-revolution and exhaust during the other. The cylinder faces are kept up against the valve face by a subsidiary hydraulic piston attached to one of the cylinders. The horizontal cylinders, which vary the position of the central stud, have a stroke of $4\frac{1}{2}$ ins. on either

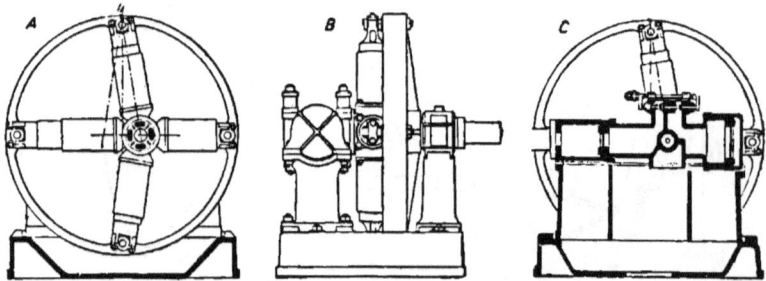

Fig. 22.

side of the centre of the crank disc. If the stud is placed con-centric with the crank-disc, the driving plungers in the four working cylinders have no motion relatively to the cylinders, and the engine can do no work. If the stud is moved eccen-trically to the right, the engine rotates in one direction. If it is moved to the left, the engine revolves in the reverse direction. The water is supplied by a pipe to the left hand cylinder in fig. 22c, and thence through the hollow plunger it reaches the in-let ports of the working cylinders. The water pressure in the hollow ram tends to move it to the right. But water pressure can also be admitted behind the larger plunger in the right hand cylinder. Then the plunger moves to the left. By two conical valves pressure is admitted to, or released from, the back of the larger plunger, so that it moves to any required position. If both conical valves are closed, the plunger is locked in position. The valves are worked by hand, and the two cylinders and plunger form an hydraulic relay. Lubrication is effected by an oil pump.

A small engine of this construction with $2\frac{1}{4}$ in. pistons, making 500 revolutions per minute, and working with a pressure of 700 lbs. per sq. in., has been driving a three ton capstan at Millwall for the last five years. There seems no reason why these engines should not work with a high efficiency, but no exact experiments on their efficiency have been published.

<h2 style="text-align:center">TURBINES</h2>

The turbine is now used in almost all cases, where the use of a pressure engine is not dictated by the nature of the work to be done, or, in the case of small motors for high falls, where the speed of rotation of a turbine would be inconvenient. In the design of turbines, theory and practice have been happily united, and every application of motors of this kind ought to be to some extent a special study. The use of such motors is of peculiar importance in connection with the distribution of power. It is where cheap water power utilised by means of turbines is available that a system of distributing power can be most safely undertaken.

Water is conveyed to a turbine, from a pentrough at the top of the fall, in which there is usually a strainer to keep out solid bodies carried by the water, in a closed supply pipe. The turbine consists essentially of a fixed casing and a revolving wheel. In the fixed casing guide passages are formed, passing through which the water acquires a definite velocity, and from which it is discharged into the wheel at a definite angle with the circumference. The function of the revolving wheel with its curved passages is to receive the water without shock, and to deviate it gradually, so that it expends its energy in driving the wheel, and is discharged with only a small fraction of energy left into the tail-race. The angular impulse or moment of the pressure driving the wheel is equal to the change of moment of momentum per second of the water flowing through the wheel. Added to the essential elements, the guide passages and revolving wheel, there are arrangements for controlling the quantity of water used, and sometimes for regulating the speed.

According to the general direction in which water flows through the wheel there are radial (inward or outward) flow turbines, and axial flow turbines. In some turbines the water flows partly radially, partly axially, and these may be termed mixed flow turbines. No essential difference of action depends on the direction of flow. In this respect the choice of a turbine depends on considerations of secondary importance, and on questions of convenience of construction.

A much more fundamental difference arises according as there is or is not a pressure in the clearance space between the guide blades and wheel. In all the older types of turbine, the water left the guide passages with part only of its pressure head converted into kinetic energy. The velocity at the inlet surface of the wheel was that due to a part only of the whole fall. Such turbines are termed *pressure turbines*. It is a condition of their proper operation that all the wheel passages should remain continuously filled with water. If the passages are alternately filled and emptied, the required pressure and velocity conditions cannot be maintained. Consequently, for a pressure turbine to act with full efficiency, the water must be supplied simultaneously to the whole circumference of the wheel.

The other fundamental type of turbine is one in which there is no pressure in the clearance space (except of course atmospheric pressure). The water issues from the guide passages with the full velocity due to the whole fall. Such turbines are termed *action* or *impulse turbines*. In such turbines the water is freely deviated by the curved vanes of the wheel, and it is best that the wheel passages should not be filled by the water. Generally air or ventilating holes are provided to prevent the filling of the wheel passages. In such wheels, since each particle of water acts independently, it is not necessary that the water should be supplied to the whole circumference. Impulse wheels have often partial admission. Then the diameter of the wheel is determined independently of hydraulic considerations. The circumferential speed being fixed with reference to the velocity of the water, the diameter may be so chosen as to give any required number of rotations per minute.

Pressure turbines work best when drowned, so that there is no lost fall due to the drop from the wheel into the tail-race; or they may be placed anywhere not more than 25 feet above the tail-race, the lower part of the fall being utilised by suction pipes, or what Americans call draught tubes. Impulse turbines cannot be drowned, for then the free deviation of the water on the wheel vanes is interfered with by the tail water in the wheel. When the tail-water level is variable, pressure turbines have an advantage, because the impulse turbine must be placed above the highest tail-water level, and some of the fall is wasted.

A compromise may be effected. The turbine may be designed so that there is no pressure in the clearance space, and

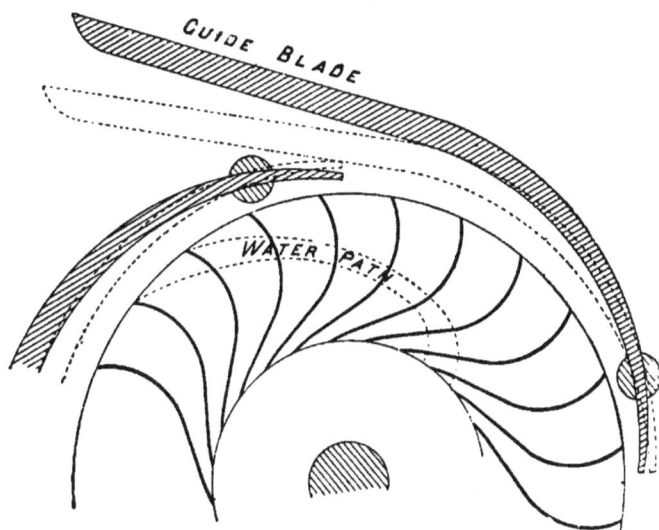

Fig. 23.

the wheel passages may have exactly the form of a freely deviated stream. Then the turbine acts normally as an impulse turbine, though the wheel passages are filled. If the wheel becomes drowned, it acts as a pressure turbine with a small pressure in the clearance space. Such a turbine is termed a *limit turbine*.

Fig. 23 shows the guide passages and wheel vanes of an inward flow pressure turbine. The curved dotted strip in the wheel is the absolute path of the water in the wheel. The water enters the wheel in a direction making an angle of about 10° with the circumference, and is gradually deviated till it leaves the wheel radially.

Fig. 24 shows similarly part of the guide passages and wheel of an axial flow impulse turbine.

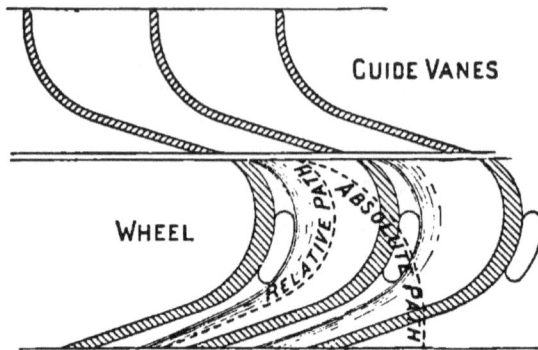

FIG. 24.

Regulation of Turbines.—The regulation of turbines for a varying load has presented considerable difficulty. For pressure turbines three modes of regulation have been tried:—(1) By means of some form of equilibrium sluice in the supply pipe the effective head may be varied to suit the work to be done. In this mode of regulation the power of the turbine is diminished, partly by diminishing the flow, partly by destroying some of the available head. It is therefore uneconomical. Generally the speed of the turbine has to be kept constant. Then at light loads the turbine has to be run at a speed above that most suitable for the balance of the head left, after deducting the resistance of the sluice. This further diminishes the efficiency. (2) Part of the guide blade passages may be closed, and the turbine worked with partial admission. This is the commonest mode of regula-

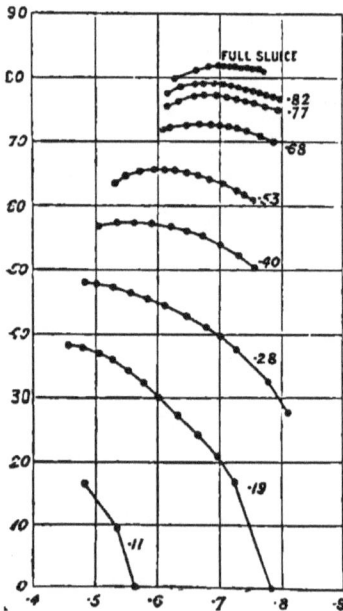

FIG. 25.

tion, because it can be effected by simple mechanical means. But it is uneconomical, because the wheel passages alternately come opposite open and closed guide passages. There is alternately flow through the wheel passages, and the water in them is arrested. Large losses due to eddying and imperfect action of the water on the wheel vanes are unavoidable. Fig. 25 shows the efficiency of a turbine with about as good a regulation of this kind as is possible. The tests were made at Holyoke.

The numbers marked against each curve are the fractions of the total guide passage area open. It will be seen that the efficiency falls off rapidly as the guide passages are closed and the wheel acts at ' part gate.' Further, the speed of greatest efficiency gets lower as the sluices are closed. Hence for a turbine working at constant speed the efficiency falls off faster than if the speed could be varied.

Thus, if in the case shown in fig. 25 the turbine had constantly to run at a circumferential speed of $0.7\sqrt{2g\text{H}}$, the efficiencies would be as follows :

| Sluice opening | . | . | 1·0 | ·82 | ·77 | ·53 | ·40 | ·28 | ·19 |
| Efficiency | . | . | ·82 | ·79 | ·77 | ·64 | ·55 | ·40 | ·20 |

These may be taken to be good results for a pressure turbine.

(3) The only mode of regulation of a pressure turbine free from obvious objections is the varying of all the guide passages in area simultaneously. That can only be done in certain types of turbine. Fig. 23 shows a plan adopted first by the late Professor James Thomson. The guide blades are pivoted near their inner ends. By moving all the guide blades together, as indicated by the dotted guide blade, the passages are equally diminished in area without much altering the direction in which the water enters the wheel.

In the case of impulse turbines the regulation is much easier, and it is for this reason that impulse turbines have been steadily superseding pressure turbines where economy of water is important. Since the efficiency of impulse turbines is not affected by resorting to partial admission, guide blade passages may be closed without much reducing the efficiency.

Professor Zeuner made tests of two Girard impulse turbines

of 200 h.p. on 12 feet fall, on Prince Bismarck's estate at Varzin. The turbine was guaranteed by the makers to give 75 per cent. with full sluice and 70 per cent. with half sluice. The results of the tests showed the efficiency to be 0·795 with full sluice and 0·801 with sluices half closed, the speed being the same in both cases.

TABLE OF CLASSIFICATION OF TURBINES

I. Impulse Turbines	*II. Pressure Turbines*
(Wheel passages not filled.)	(Wheel passages filled.)
Discharge always above the tail water.	Discharge above or below the tail water or into a suction pipe.
1. With complete admission for full load.	Always with complete admission for full load.
2. With partial admission for full load.	
Regulation. Usually by closing part of the guide passages.	Regulation. (*a*) by throttling; (*b*) by closing some of the guide passages; (*c*) by varying the area of the guide passages.

(*a*) Radial flow, inward.
(*β*)   „   „   outward.
(*γ*) Axial flow.
Mixed flow turbines are pressure turbines only.

The Efficiency of Turbines.—The efficiency of turbines working with full load and designed with a view to the best results is generally taken for calculation at 0·75 to 0·80, but many turbines of good construction when tested have given at their best speed efficiencies ranging from 0·80 to 0·85. The losses vary in different cases, and cannot be precisely assigned by any general equation. On the average they have about the following values in per cent. of the effective work of the fall :—

	Per cent.
Shaft friction and leakage	3 to 5
Unutilised energy	3 to 7
Friction and shock in guide and wheel passages	10 to 15
Total	16 to 27

Choice of Type of Turbine.—The first general condition which determines the choice of the type of turbine is the variability or constancy of the tail-water level. Since impulse turbines will not work drowned, it is necessary to choose a pressure turbine or a limit turbine when the tail water level is liable to rise in flood.

The next general consideration is that the regulating apparatus of impulse turbines is simpler and more efficient than that
of pressure turbines. Hence when the water supply is variable
and scanty the impulse turbine is generally to be preferred.

The third general consideration is that on very high falls
the rotational speed of pressure turbines is inconveniently great.
In such cases a partial-admission impulse turbine is preferable,
because it permits an increase of the wheel diameter and consequently a decrease of the speed of rotation.

When there is an ample supply of water at all times the
pressure turbine is generally adopted, unless its speed is too
great. It can be constructed more cheaply than an impulse
turbine. On moderate falls with variable water supply and varying tail-water level the limit turbine is better than the pressure
turbine, because it has a higher average efficiency. For very
high falls the impulse partial-admission turbine is generally
best, on account of its low speed of rotation and the simplicty
of its regulating apparatus.

It is a popular delusion that by some lucky trick or specialty
of design a turbine can be made better than that of other
makers. The principles of turbine-designing are well determined, and the various patents taken out for turbines involve
only details of manufacture, and not anything essential to
efficient working.

Speed-governors for Turbines.—For some time the question
of automatically regulating the speed of turbines was not of
great importance. With large low-pressure turbines on a fairly
constant fall, hand regulation was sufficient. Some forms of
turbine are naturally stable in speed. Now that high falls are
utilised, and especially since turbines have been applied to drive
electrical machinery, automatic regulation has become much
more essential.

The regulation of the water supply to a turbine is a more
difficult problem than the regulation of the steam supply to an
engine. Turbine sluices are heavy, and require considerable
power to move them. In consequence of the momentum of the
column of water in the supply pipe the regulation must be
gradual. Ordinary governors are too feeble to act directly on
water sluices. A relay must be used to work the sluices put in
action by a governor.

In the case of some turbines a pendulum governor puts in action one or other of two clutches which gear the sluices to the turbine itself. Governing in this way is never satisfactory. The action of the relay lags behind the action of the governor. Suppose part of the load thrown off. The speed increases and the governor engages the clutch which closes the sluices. But the action is slow, and a considerable excess of speed is at-

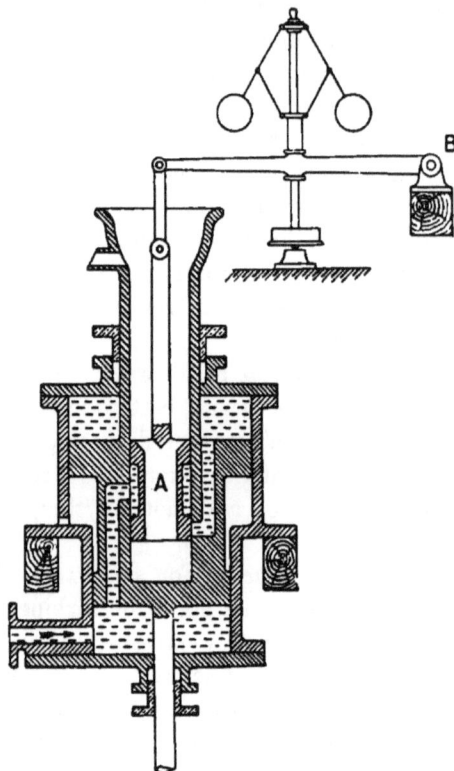

Fig. 26.

tained before the closing of the sluice stops the increase of speed. The closing must then still continue till the speed begins to fall. But when the speed begins to fall the effort of the turbine is no longer equal to the resistance of the driven machines. The sluice has been closed too much. The reverse action then begins, the action of the sluice being again too slow to check the decrease of speed till it has fallen below the normal. A periodic fluctuation of speed is so set up.

A fly-wheel which absorbs or restores part of the work modifies this, and gives the governor relay more time to make an adjustment. Messrs. Rieter have used a fluid brake to absorb part of the excess work with a similar object.

For the impulse turbines in Geneva Messrs. Faesch & Piccard have used the relay governor shown in fig. 26. The governor acts directly on the piston valve, A, of a hydraulic relay cylinder. In the cylinder is a piston with two unequal faces. On the lower and smaller face there is a constant water pressure tending to raise the piston if the water in the upper chamber is not locked by the valve. If the governor balls rise the valve allows water to flow from the upper chamber, and the piston rises. Conversely, if the governor falls the valve admits pressure water from the lower chamber to the upper chamber, and, the upper piston being the larger, the piston descends. The movement of the ram tends to close whichever port has been opened, so that unless the governor continues to rise the action on the turbine sluices is stopped. In that way the sluices are operated as if they were directly connected to the governor. For each position of the governor there is a definitely fixed corresponding position of the sluice.

This governor answers perfectly for impulse turbines with small sluices, because the whole range of action of the sluice can be obtained with a small alteration of height of a sensitive governor, corresponding to a small variation of turbine speed. But there will be a variation of speed corresponding to the range of governor height necessary to give the required movement of the sluice, which is virtually attached rigidly to it. In some later governors Messrs. Faesch & Piccard have introduced a correcting adjustment. The movement of the piston adjusts the height of the fulcrum B, raising it when the governor is rising and lowering it when the governor is falling. The range of action of the sluice is then made independent of any variation of governor height.

CHAPTER VII

TELODYNAMIC TRANSMISSION

THE origin of the first system by which power was transmitted to distances which, at least at the time, appeared considerable is interesting.[1] In 1850 there were at Logelbach, near Colmar, in Alsace, some buildings which had formed the manufactory of printed calicos of MM. Haussmann. This factory, founded in 1772, had been reduced to inaction since 1841 by the decay of its ancient industry. A plan was sought for restarting the factory as a weaving factory. There was only one steam engine, and the buildings were scattered at considerable distances. To use shafting for transmitting the power would have involved too much expense and loss. It occurred to M. C. F. Hirn to drive one of the buildings at a distance of 80 yards from the engine by a steel band 2 ins. wide and one twenty-fifth of an inch thick, on wooden pulleys 6½ feet in diameter and making 120 revolutions per minute.

The band was not entirely successful. The least wind put it in vibration and the pulley guides tore it at the riveted joint. It worked however for 18 months, transmitting 12 h.p. Then an English engineer, Mr. A. Tregoning, suggested the use of a wire cable. A wire rope, furnished by Messrs Newall & Co., ¼ inch in diameter was substituted for the band. The same pulleys were used with a groove ½ inch deep turned in the rim. That cable worked for many years, iron pulleys having been substituted for the wooden ones. A second transmission to a distance of 256 yards with a cable ½ inch in diameter on pulleys 9½ feet in diameter, running at 92½ revolutions, and transmitting

---

[1] *Notice sur la Transmission télodynamique*, par C. F. Hirn ; Colmar, 1862. *Note sur la Transmission télodynamique inventée par M. C. F. Hirn*, par M. du Pré Bruxelles, 1869. *Erfahrungsresultate über Betrieb und Instandhaltung des Drahtseiltreibs ;* Ziegler, Winterthur, 1871.

50 h.p., was soon erected. Supporting pulleys were used at the half-distance. The author saw this transmission still in use in the present year.

M. Hirn has stated that his chief difficulty was the construction of the pulleys. It was not till he tried a pulley having a dovetailed groove, filled with gutta percha as a seating for the cable, that he felt the problem of telodynamic transmission to be solved. The durability of the transmission was of the first importance from the expense and trouble of replacing the rope. With these pulleys neither pulley nor cable suffered excessively from wear.

The amount of work transmitted by a cable is proportional to the product of the effective tension (difference of the tensions in the tight and slack sides) and the speed. To transmit power to great distances by manageable cables, the strongest material must be used for the cables, and they must be run at the highest practicable speed. The cables were at first of iron, now they are generally of steel. The largest cables which it appears to be practicable to use are about one inch in diameter. In order that the bending stress may not be excessive the pulleys are of large diameter, 12 feet to 15 feet usually. For the throat of the pulley on which the rope runs gutta percha, soft wood, and leather have been used. At present the bottom of the pulley groove is usually formed of strips of waste leather forced into a dovetailed groove. The greatest allowable speed of rope is that at which the centrifugal tension of the pulley rims becomes dangerous. One hundred feet per second has been adopted as the greatest practicable speed. The pulleys are placed at maximum distances of 300 to 500 feet apart. The weight of the rope then ensures sufficient adhesion to prevent slipping, when the ropes are tightened so that the deflection or sag is not inconvenient. With these limitations, a 1 inch rope will transmit about 330 h.p.

Soon after the erection of the transmissions at Logelbach M. Henri Schlumberger transmitted the power of a turbine 86 yards to work agricultural machinery. In 1857, at Copenhagen, Captain Jagd transmitted 45 h.p. to saw-mills at a distance of more than 1,000 yards. In 1858, at Cornimont, in the Vosges, 50 h.p. was transmitted 1,251 yards. In 1859, at Oberursel, 100 h.p. was transmitted 1,076 yards; and at Emmendingen 60 h.p. was transmitted 1,312 yards.

In 1862 Hirn stated that about 400 applications of the telodynamic system had been constructed by Messrs. Stein & Co., of Mulhouse, carrying an aggregate of 4,200 h.p. over distances amounting altogether to 80,000 yards. Some later transmissions have been constructed by Messrs. Escher Wyss, of Zürich, and by Messrs Rieter Brothers, of Winterthur. Hirn wrote in 1862, with pardonable exultation, that 'jusqu'à présent la force motrice était localisée. désormais elle sera mobilisée.'

General Description of the System of Transmission by Wire Rope Cable.—The cables used are stranded ropes having 6 to 12 wires in each strand and 6 to 10 strands in each rope (fig. 27). The strands are twisted on a hemp core, and usually there is a hemp core to each strand. The hemp makes the ropes flexible. At first Swedish charcoal iron was used, now the ropes are more commonly of steel. They are protected from oxidation by a

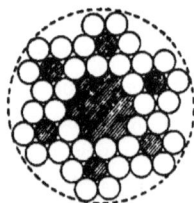

FIG. 27.

coating of boiled oil. It is necessary to keep a spare rope in reserve in case of accident.

The ratio of the tensions in the tight and slack sides is about 2 to 1. In passing round the pulleys the rope is bent, and the bending stress is added to the longitudinal stress. In crder that the bending stress may not be excessive large pulleys must be used. Their diameter is so chosen that the bending stress and longitudinal stress are about equal. As the tension in the slack side is less than that in the tight side, greater bending stress may be permitted, and consequently supporting pulleys for the slack side may be smaller than the driving pulleys.

When first erected the rope stretches a good deal from the tension compressing the hemp cores and diminishing the twist of the rope. The rope must then be re-spliced, which is troublesome and costly. To diminish this difficulty Messrs. Rieter Brothers pass the rope before use between grooved rollers, which compress it laterally. This consolidates the rope. In 10 to 15 passages through the rollers it stretches from 1 to 4 per cent. in length, and it diminishes in diameter about 6 per cent.

The following table gives data of some of the most extensive cable transmissions :—

TABLE OF TELODYNAMIC TRANSMISSIONS

Place	Total h.p. transmitted	H.P. on one rope	Total distance of transmission	Diameter of pulleys in ins.	Velocity of rope in ft. per second	Diameter of rope, ins.	Diameter of wire, ins.	No. of strands	No. of wires in each strand	Total No. of wires
			Ft.							
Oberursel	104	104	3,170	148	73	·63	·07	—	—	48
Schaffhausen	560	280	—	177	62	1·08	·072	8	10	80
„	—	150	—	177	—	·88	—	—	—	60
Freiberg	300	300	2,510	177	62	1·08	·072	10	9	90
„	—	50	—	108	—	·64	·043	8	9	72
„	—	120	—	148	—	·72	·060	6	11	66
„	—	60	—	148	—	·48	·054	6	6	36
„	—	20	—	84	—	·36	·036	7	6	42
Bellegarde	3,150	300	—	216	65	1·28	·088	8	9	72
Tortona	8	8	—	78	—	·43	·039	6	8	48
Zürich	—	150	—	186 and 127	—	·80	—	—	—	—

The total distance of transmission at Freiberg is 6,500 feet.

Supporting Piers.—The piers supporting the pulleys are expensive, and each pulley over which the rope passes is itself a source of wear and frictional loss of power. It is desirable, therefore, that the spans of the cable should be as large as possible. Spans of 300 to 500 feet have been commonly adopted. For such spans the deflection of the rope is considerable, and must be provided for by giving sufficient height to the piers. The deflection of the slack side of the rope is greatest, and hence commonly the slack side is placed above the tight side.

Arrangement of Spans.—Fig. 28 (I), shows a single span with the deflections of the tight and slack sides of the rope. At (II) is shown a single span, with one supporting pulley for the slack side of the rope. For greater distances of transmission a single rope may still be used as at (III), with supporting pulleys for both tight and slack sides of the rope. M. Ziegler, the engineer of Messrs. Rieter Brothers, first constructed a transmission with a series of independent spans, fig. 28 (IV). Then the intermediate pulleys are two-grooved. Professor Reuleaux, of Berlin, has proposed the arrangement shown in fig. 28 (v) to reduce the height of the supporting piers. The spans of the tight side are twice as long as those of the slack side, so that the deflections are nearly equal, and the supporting pulleys for the tight side are twice the diameter of those for the slack side,

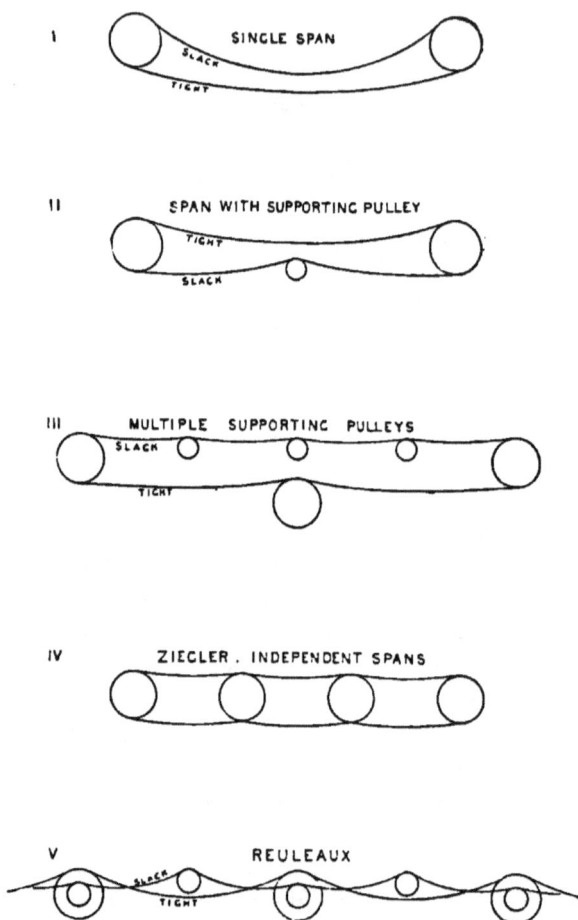

FIG. 28.

so that the total longitudinal and bending stress is about equal also.

To give the rope a good frictional hold on the pulley, and to increase its durability, the pulley grooves are bottomed with some softer material than iron. Fig. 29 shows a section of single and double grooved pulley rims, with the leather strips in a dovetailed groove, which are now generally used. The pulleys are of cast iron, or of cast iron with wrought-iron arms.

As the principal loss of work in transmission is due to the weight of the pulleys and rope resting on the journals, the

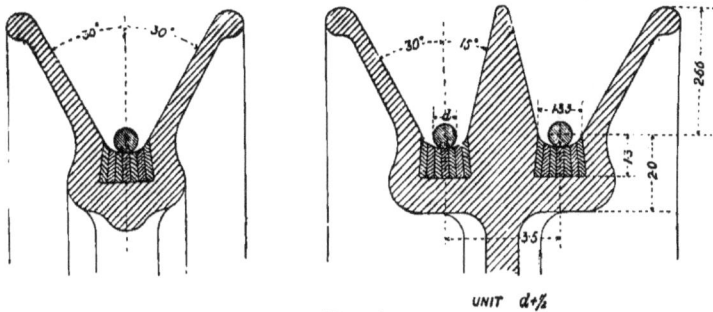

Fig. 29.

pulleys should be made as light as possible. The leather packing in the grooves requires renewal every three or four months, and this involves a not inconsiderable cost in maintenance.

The pulley stations must be substantially constructed. They are sometimes of timber, sometimes of iron or masonry. Stations at which the direction of the transmission changes are called change stations. The change of direction has been

Fig. 30.

generally effected by bevil gearing. Fig. 31 shows one of the masonry pulley stations of the Bellegarde transmission. Fig. 30 is a change station with bevil gearing. Reuleaux proposes to change the direction of the ropes by guide pulleys, as shown in fig. 32.

Efficiency of Cable Transmission.—The two principal sources of waste of work in cable transmission are the friction of the

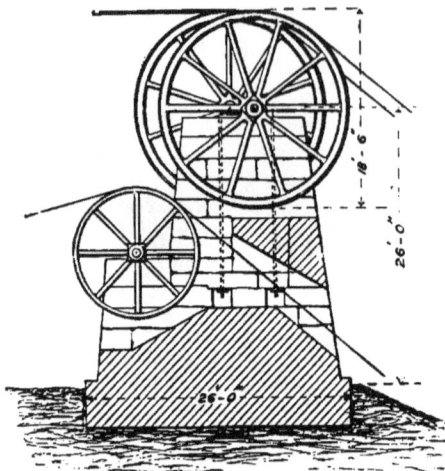

FIG. 31.

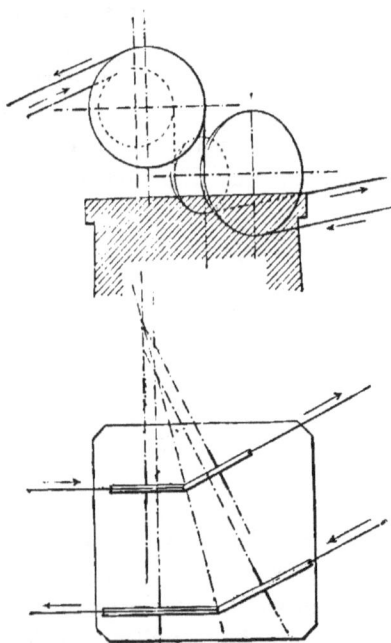

FIG 32.

journals supporting the pulleys, and the resistance to bending of the rope at the points where it comes on and off the pulleys. Ziegler made experiments at Oberursel on a transmission of seven spans transmitting 104 h.p. The total loss of work at eight stations was $13\frac{1}{2}$ h.p., or, say, 1·7 h.p. at each station. When transmitting full power, the efficiency of the system is remarkably high. Probably for moderate distances the efficiency is greater than for any other mode of transmission, except electrical transmission. But the waste of work is the same for all loads transmitted, so that when working at less than full power the efficiency falls off. If from Ziegler's experiments the efficiency for a single span (two pulley stations) is taken at 0·967, then for a transmission of m intermediate and two terminal stations the efficiency, working with full load, is—

$$\eta = 0·967^{\frac{m+2}{2}}$$

Number of spans	.	.	.	1	2	3	5	7	9
Number of pulley stations		.	2	3	4	6	8	10	
Efficiency .	.	.	.	0·967	·951	·934	·904	·874	·845

At half load, however, the loss of work for one span would
be the same as for full load, so that the efficiency of one span
would be only 0·934. Consequently the efficiency for several
spans would be as follows :—

Number of spans	.	.	.	1	2	3	5	7	9
Number of pulley stations		.	2	3	4	6	8	10	
Efficiency .	.	.	.	·934	·902	·872	·815	·761	·711

The Oberursel Transmission.—Not long after the invention
of Hirn's system of wire-rope transmission, and its application
at Logelbach, an opportunity occurred for trying it on a more
considerable scale. A cotton mill had been built at Oberursel,
near Frankfort, to utilise the water power of Urselbach. Two
tangential wheels were erected on a fall of 165 feet. Their
total power varied from 64 to 150 h.p., according to the condition of the stream. In 1860 more power was required. A
fall of 264 feet was found above the mill, but at a distance such
that it could not have been made available by ordinary means
of transmission. On this fall two tangential wheels were
erected, each yielding from 40 to 104 h.p., according to the
condition of the stream. The water from the upper turbines
afterwards drives the lower turbines. From the upper turbines
the power is transmitted by wire-rope cable a distance of 3,160
feet, in seven spans of about 400 feet.

Leather lining was first adopted for the pulleys of this transmission. The pulleys are 12½ feet in diameter, and make 114½
turns per minute, the rope having a speed of 73 feet per second.
The pulleys on intermediate stations are double-grooved, and
there is a separate rope to each span. The cables are ⅝ inch in
diameter, having 6 strands of 6 wires 0·06 inch in diameter.
The tensions at full load are about 11,400 lbs. per sq. in. due
to bending, and 14,200 lbs. per sq. in. due to longitudinal tension,
or 25,600 lbs. per sq. in. altogether. The inclination of the spans
increases this tension a little. The installation was designed by
M. Ziegler, and carried out by Messrs. Rieter Brothers, of
Winterthur.

The Ochta Transmission.—In 1864, after a serious explosion, the gunpowder factory at Ochta near St. Petersburg was rebuilt, and wire-rope transmission was adopted in order to secure the condition that the buildings should be at a safe distance from each other. The motive power of the new factory was supplied by two turbines of 140 h.p. each, and a similar turbine in reserve. The buildings were erected in three lines, one through the principal axis of the turbine-house, another parallel to this at a distance of 420 feet, and a third at right angles to it, and passing through the shorter axis of the turbine-house.

In each line the buildings nearest the turbine-house were 330 feet from it. In the first line there were eight buildings, requiring 100 h.p., and the buildings were 164 feet apart centre to centre. The second line contained twelve buildings placed 230 to 330 feet apart, and requiring 80 h.p. The third line contained three buildings, requiring 24 h.p. and placed about 300 feet apart. There were therefore twenty-three buildings widely scattered to be supplied with power. The greatest lengths of transmission were 1,300, 2,300, and 2,600 feet. The power was transmitted by wire-rope cables, the work being carried out by M. Stein, of Mülhausen.

The Schaffhausen Transmission.—A still more considerable application of telodynamic transmission, which attracted general attention, was made soon after at Schaffhausen.[1] After a period of trade depression there was a revival of industry at Schaffhausen between 1840 and 1850. In the year 1850, Herr Heinrich Moser, of Charlottenfels, constructed a canal and erected the first turbine at Schaffhausen. It then occurred to him that it might be possible to render useful the immense volume of water passing down the rapids of the Rhine in front of the town. An extraordinary low condition of the river, in the winter of 1857-8, favoured an examination of the bed of the river, and a Commission was appointed to mature a project. This Commission suggested the formation of a weir across the river; the construction of a water-power station in the bed of the river near the left bank, where the conditions were suitable

---

[1] *Turbineanlage und Seiltransmission der Wasserwerkgesellschaft in Schaff-hausen,* von J. H. Kronauer; Winterthur, 1870. *Die Wasserwerkgesellschaft in Schaffhausen ;* Schaffhausen, 1889. *Fünfundzwanzigster Jahresbericht des Verwaltungsrathes der Wasserwerkgesellschaft in Schaffhausen,* 1889; Schaff-hausen, 1890.

for excavation; the erection of turbines of 500 h.p. on a fall of 12 to 15 feet; and the transmission of the power across the river to the factories by the then new system of telodynamic transmission. The cost was estimated at 20,000*l.*

Fig. 33 shows the general arrangement of the works at Schaffhausen, with the old turbine-house for the cable transmission and the new power house for electric transmission, which will be described later.

A company was formed in 1864 to carry out the works. A weir was constructed during favourable seasons in 1864–6, across the rocky bed of the river, which is about 500 feet wide. The fall immediately at the weir is not great, but there are rapids below it. By placing the turbine-house in the river-bed near the weir and constructing a tunnel tail-race 620 feet in length, a fall was obtained which varies from 15·6 to 13·7 feet. The turbine-house contains three axial flow pressure turbines with vertical shafts of 200, 260, and 300 h.p., or 760 h.p. altogether. They gear with a common horizontal shaft by means of bevil wheels. Any turbine can be put out of gear by lowering the bevil wheel on the vertical shaft.

Fig. 34 shows a section of the turbine-house, with the head

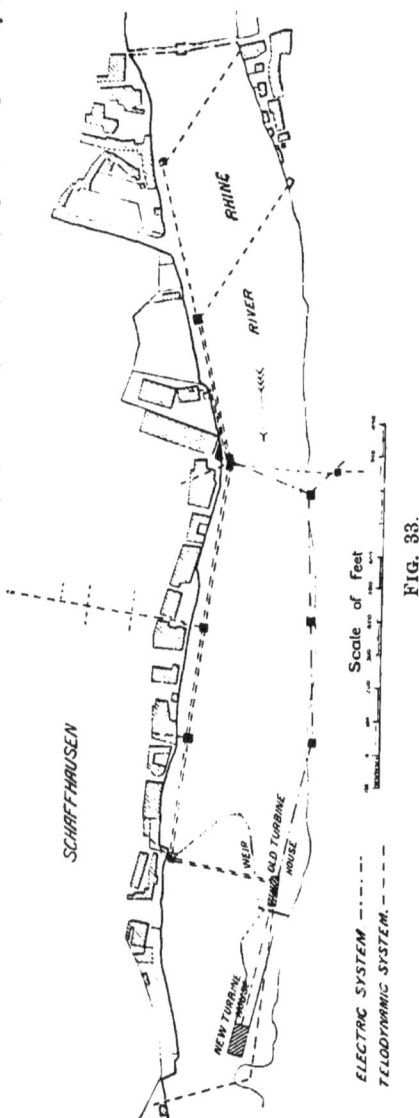

gates, the gearing, and the pulleys of the principal cable transmission.

About 150 h.p. is transmitted from one of the turbines to a
factory on the hill above the turbine-house, by a steel shaft 550
feet in length. From the same shaft also about 22 h.p. is transmitted, by a small cable passing down the left bank of the river
and then crossing it, to a pulp factory on the right bank. This
leaves a maximum of about 570 h.p. to be dealt with by the
main cable transmission, which crosses the river directly from
the turbine-house, and then passes along the right bank to the
factories, as shown in fig. 33.

As to the turbines, there is nothing of special interest except
that they are constructed with a partition dividing them into
two rings of buckets. During low conditions of the river,
when there is a good fall, the outer ring of buckets only is
used. When the fall is smaller both rings are used. This
compensates a little for the variation of normal or most economical speed of periphery with variation of fall. The turbines
make 34·28 revolutions per minute. The turbine regulating
sluices are under the control of a relay governor. An auxiliary
6 h.p. turbine works the main sluices of the inlet chamber.
There is a friction brake on the main shaft of the turbines,
which is thrown into action by the governor if the speed
exceeds a certain limit. Then, if a rope breaks, the friction
brake comes into action and stops the turbines. Connected
with the brake is an apparatus for determining the power
transmitted by the ropes; but the author has not been able to
learn whether this is satisfactorily used.

It has been seen that the turbines drive by bevil wheels a
horizontal shaft, each turbine being capable of being disconnected
by putting its bevil pinion out of gear. The horizontal shaft
makes 80 revolutions per minute, and carries at its driving end
two principal rope pulleys of 14·75 feet in diameter, as shown in
fig. 34. From these pulleys two cables cross the river in a single
span of 385 feet to a pulley station in the river at the left bank,
where the direction of the transmission is changed by bevil gearing, and thence the transmission passes up the left bank of the
river. The two principal rope-driving pulleys are not keyed
on the horizontal driving-shaft, but run loose on it. Between
them is a strong cross-head keyed on the shaft, carrying bevil

wheels on studs, which gear into bevil wheels fixed to the driving
pulleys. If the tension in the two driving cables is the same,
the bevil gearing would have no action. But if there is any

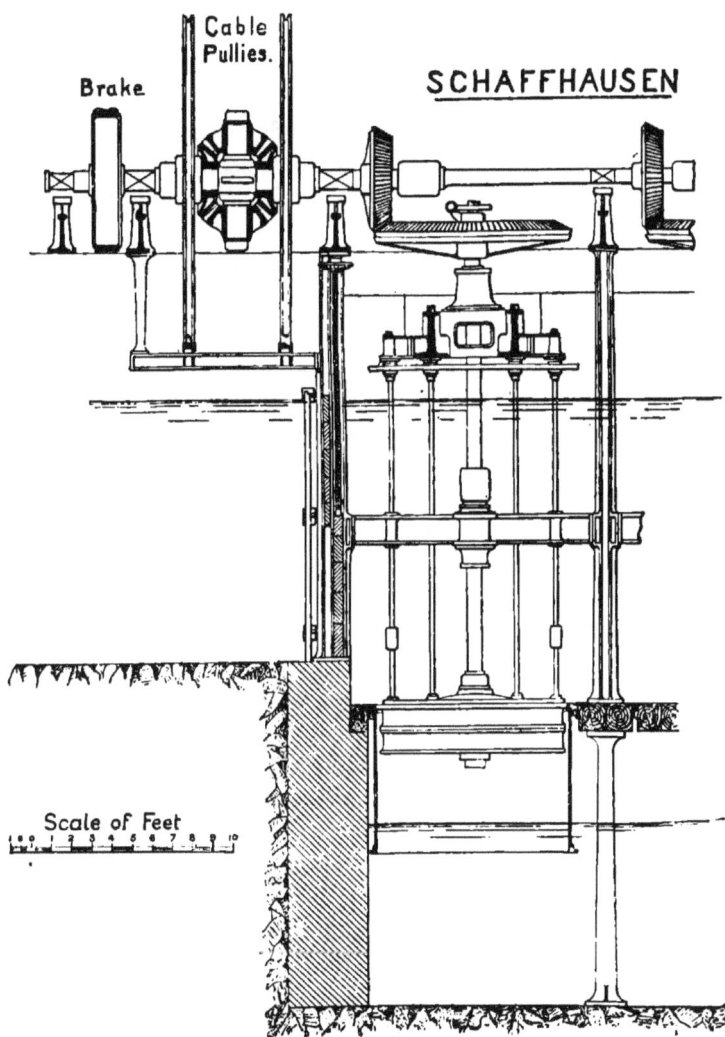

Brake

Cable
Pullies.

SCHAFFHAUSEN

Scale of Feet

FIG. 34.

difference of tension the bevil wheels permit one driving pulley
to rotate faster than the other, so that equality of tension in the
ropes is re-established. This differential gear has not so far as

the author is aware been elsewhere adopted. Provision is made
for keying either driving pulley on the shaft in the event of
one rope breaking and the power having to be transmitted
through the other rope. The differential gear is then out of
action.

The gross power in the horizontal driving shaft in the
turbine-house is about 530 h.p., or, allowing for friction, say 500
effective h.p., to be transmitted to the factories, or 250 h.p. for
each rope. Either rope is capable of transmitting at any rate
a large fraction of the whole power temporarily, if the other
rope is broken.

This power is delivered by the ropes at the change station
on the left bank. At that station about 22 h.p. is taken off
by prolonging the second shaft of the bevil gearing and a subsidiary rope transmission. The remaining 478 h.p. is transmitted along the left bank to the first intermediate pulley station
at a distance of 370 feet by a pair of cables. Thence to the
second intermediate station, distant 345 feet, by another pair of
cables. At 455 feet further is a second change station, at which
the direction is again changed by gearing. Thence the ropes
pass to two other intermediate stations.

From the second intermediate station an underground shaft
carries about 27 h.p. to ten small workshops, and from the
second change station, and the third and fourth intermediate
stations, cables are carried back across the river to factories on
the right bank. From the first shaft at the second change
station about 110 h.p. are distributed, partly by a special rope
gear, partly by vertical and underground shafting, to four
factories, one of which is the large Mosersche Gebaude ; and
from the second shaft of this station a steel shaft transmits
200 h.p. to Scholler's wool factory.

Between the turbine-house and the second change station
the cables consist of 80 wires, 0·042 inch in diameter, in 8
strands of 10 wires each, with a hemp core. Diameter of rope,
1·08 inch. Speed of rope, 62 feet per second. Smaller ropes
are used in other parts of the transmission. The distribution
has been described rather fully, because it is essential to learn
how far wire-rope transmission can be adapted to complex conditions where many consumers require power.

It would appear that the rather complex arrangement of

differential gear and double cables was intended originally to meet the case of having to drive by a single cable, while the other was broken or under repair. It appears that under present conditions one cable would not be strong enough to transmit the whole power which is utilised. On the other hand, experience has shown that there is no necessity to duplicate the cable to avoid accidental stoppages. The total length of principal transmission is about 2,000 feet.

The Schaffhausen installation has been an entirely successful undertaking, and has very greatly benefited the industries of the town. Some particulars of the extent to which the power has been utilised may be interesting :—

Year	Number of renters of power	Average total h.p. supplied	Rent from power
			£
1867	13	121	345
1868	13	150	585
1869	14	180	730
1870	15	254	830
1871	16	293	1,215
1872	17	337	1,510
1873	17	350	1,650
1874	19	483	2,300
1875	21	527	2,520
1876	23	595	2,800
1877	23	597	3,000
1878	23	610	3,060
1879	23	610	3,040
1880	24	623	3,120
1881	23	634	3,100
1882	23	641	3,200
1883	23	639	3,240
1884	22	641	3,080
1885	23	632	2,880
1886	23	639	3,080
1887	23	641	3,300

The charge for power, in 1887, varied from 4*l.* 16*s.* to 6*l.* per h.p. per annum.

The total cost of the works appears to have been reckoned at 29,360*l.* originally, and this by writing off stood at 24,664*l.* in 1887.

As to the working of the system, it appears that experience has proved that there is a greater loss of power in transmission in wet and frosty weather than was originally expected. When the maximum power is being used there are oscillations of the ropes which drive the machines at irregular speed. The spinning factory suffered most from this, and ceased to take power. This has led to the construction of a new power station and the adoption in the new works of electrical instead of wire-rope transmission.

In some excellent lectures which were delivered at the Society of Arts, in 1891, by Mr. Gisbert Kapp, the method of transmission by wire rope is compared with the method of transmission electrically, very much to the disadvantage of the former. 'Till recently,' said Mr. Kapp, 'rope transmission held the field absolutely, not because it was perfect, but because there was nothing better. Now, however, we have something better in electrical distribution, and the flying ropes are being steadily replaced by the electric conductors.' The use of the words · steadily replaced' conveys a wrong impression. The wire ropes have not been replaced at Schaffhausen by electric cables, but an additional power station has been erected, and an electric transmission has been placed beside the rope transmission. It happened accidentally that, at a visit of the author to Schaffhausen about a year and a half since, the rope transmission was working while the electric transmission was stopped, having been temporarily disabled by a lightning accident. It seemed desirable to ascertain from Messrs. Rieter what view they took of the prospects of wire-rope transmission, looking to the fact that they had the opportunity of knowing the results of the working of rope and electric transmission side by side. They were good enough to send answers to some inquiries. They say that at Schaffhausen the rope plant is expected to do more work than was originally provided for or intended. Also, it was the first large installation of the kind, and had some defects which experience has shown can be remedied. Electric transmission, they say, has also been found to have some inconveniences. They, however, do not think that electric transmission will compete seriously with rope transmission for moderate distances, such as that at Schaffhausen, as I understand them. On the other hand, for long distances they admit that electric transmission has the advantage.

Cable Transmission at Fribourg, Switzerland.—In 1870 a company was formed, partly to acquire and work the forest

owned by the town of Fribourg, partly to carry out a scheme of
water supply, and partly to utilise and sell water power. This
was a scheme in advance of any previous one, because the company acquired land, which they proposed to lease to industrial
undertakings, including in the lease a right to a supply of
motive power from the water-power station of the company.

The Sarine, an affluent of the Aar, flows in a deep cut
channel near the town. A masonry and concrete dam, about
40 feet in height, was built across the river, so as to form a
considerable storage reservoir in the river-bed above the dam.
The turbine-house was built near the end of the dam on the
right bank. The ravine through which the river flows is not
suitable for sites for factories. The company therefore acquired some level land about 300 feet above the river, adjoining
a railway and otherwise well adapted for industrial establishments. It was intended to work factories built on this land by
power transmitted from the turbines by wire ropes.

With the minimum flow of the Sarine, and an effective fall
of 35 feet, 1,700 h.p. could be obtained. Provision was made
for two turbine-houses containing 8 turbines of 300 h.p. each.
Only two turbines have actually been constructed, one driving
pumping machinery for water supply, the other driving a cable
transmission. The turbines are Girard turbines, running at
74½ revolutions per minute. The turbine for transmission
drives a horizontal shaft at 81 revolutions per minute by bevil
wheels. This carries a 15-foot pulley with single groove, driving
the cable by which power is transmitted to the plain of Perolles.
The principal rope transmission consists of five equal spans of
500 feet each. The total distance to the saw-mill at which
power is first taken is 2,500 feet, and the difference of elevation
is 268 feet. The rope pulleys are all 15 feet in diameter, and
the rope is 1·08 inch in diameter. The rope consists of 90
wires ·072 inch in diameter, in ten strands of nine wires with
hemp cores. The speed of the rope is 62 feet per second. At
the saw-mill a subsidiary transmission works a rope tramway
incline for carrying timber, which uses 50 h.p. This has a cable
⅞ inch in diameter. From the shafting of the saw-mill another
subsidiary rope transmission takes 120 h.p. to railway carriage
works at a distance of 930 feet. From the carriage works there
is a further transmission of 60 h.p. by a ½ inch rope a distance

of 1,600 feet, and thence by ropes ¾ inch in diameter to a foundry and chemical factory. The power is sold at the rate of 8*l.* per h.p. per annum.

Cable Transmission at Bellegarde.—In 1872 a company was formed to utilize the water power of the Rhône at Bellegarde, not very far from Geneva. Phosphatic deposits occur near this point, and power was required in quarrying and grinding these minerals. It was expected also that other industries would be attracted to a site where power could be obtained. The Rhône flows between Fort de l'Ecluse and Seyssel in a winding gorge so narrow and deep at one part that in low water the river disappears. This part is termed the *Perte* du Rhône. A site for a power station was found almost in the bed of the Valserine near its junction with the Rhône. A tunnel was constructed from a point above the Perte du Rhône to the Valserine, calculated to discharge 2,120 c. ft. per second, with a mean fall of 36 feet. This would give nearly 7,000 h.p., but a part only has been utilised, and the works have not been financially as successful as was expected.

Five Jonval pressure turbines of 630 h.p. each have been erected. The power is transmitted upwards from the gorge to the plain of Bellegarde by wire cables, and is distributed to several works. There are phosphate works, a wood pulp factory and paper mill, a copper refinery and a pumping station. The power is sold at 8*l.* to 12*l.* per h.p. per annum.

The horizontal shafts driven by the turbines carry each two pulleys, 18 feet in diameter. The cables are 1·28 inch in diameter, consisting each of 72 wires, 0·088 inch in diameter, with a hemp core. The rope speed is 65 ft. per sec. The greatest span at Bellegarde is 630 feet.

The Cable Transmission at Gokak, in India.—A large telodynamic transmission has been recently erected at Gokak, in the Southern Mahratta country in India. This installation has been carried out by Messrs. Escher Wyss, of Zürich.[1]

A river falling over a high cliff has motive power enough for many industries. At present three turbines of 250 h.p. each (750 h.p. altogether) have been erected to drive a cotton mill of 20,000 spindles. The water, taken at 2,300 feet above the fall, is led by a channel to the edge of the cliff, and thence

[1] *Engineering,* June 8, 1888.

in a 32-inch wrought-iron pipe. The pipe descends about 110 feet vertically on the face of the cliff, and then is inclined at

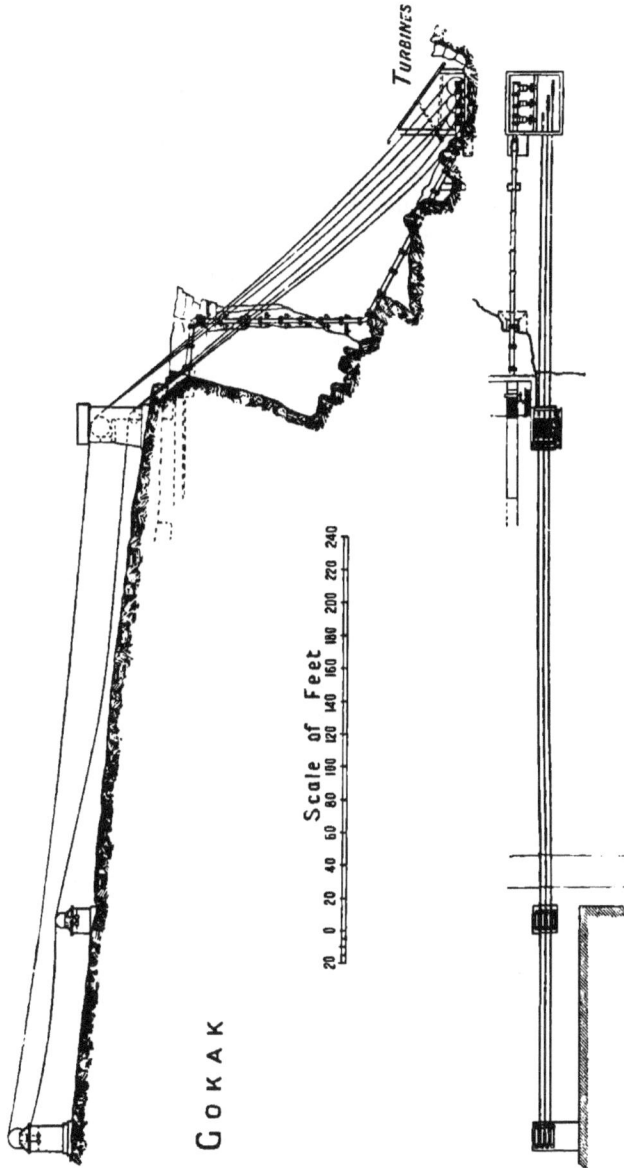

FIG. 35.

Scale of Feet

GOKAK

TURBINES

about 30° to the horizontal. In the turbine-house the three turbines are supplied by three 21-inch branch pipes. The total fall acting at the turbines is 180½ feet. The turbine wheels are 67 inches in diameter, and they run at 155 revolutions per minute.

The turbines are action or impulse turbines with partial admission, and they have horizontal axes, each turbine axis carrying a wire-rope-driving pulley. There is a sluice valve worked by hand, and a throttle or disc valve controlled by a governor to each turbine. The governor is a relay governor in which a belt on speed cones drives differential gearing connected with the throttle valve. If the governor moves the belt to either side of its central position, the differential gear comes into operation and opens or closes the throttle valve.

The shafts of the turbines carry wire-rope pulleys 11½ feet in diameter. The rope speed is 93 feet per second. The ropes are 1 inch in diameter. At the top of the cliff is a rope station, with carrier pulleys 198 and 220 feet above the turbine shafts. The pulley for the tight side of the belt is 11½ feet in diameter. That for the slack side is 8 feet in diameter. The distance from this station to the mill is 432 feet, and there is an intermediate carrier station for the slack side of the belt, which would otherwise foul the ground. There are of course 3 ropes, one to each turbine. The installation was set to work in October 1887.

Advantages and Disadvantages of the Telodynamic System.— The telodynamic system is adapted for transmitting and distributing power to distances of a mile or more, which are large compared with the distances to which power is ordinarily transmitted by shafting and similar means. On the other hand it cannot seriously compete with electrical transmission, in cases where the distance to be covered is reckoned by many miles. With this limitation it may be noted that—

(1) It has the peculiar advantage that it transmits the mechanical energy developed by the prime mover directly, without any intermediate transformation. In electrical distribution a double transformation is necessary : a transformation into electrical energy by a dynamo, and retransformation back into mechanical energy by an electric motor. This double transformation involves waste of power and increase of capital expended.

(2) The efficiency of transmission to such distances as those at Schaffhausen is undoubtedly very great. It is uncertain

whether a similar amount of power could be distributed to an equal number of consumers electrically with as little waste of energy in the process.

(3) The telodynamic transmissions which have been at work, some of them since 1864, have actually worked continuously without serious stoppage, and have only failed to return an adequate profit where they were undertaken on a scale too large for the amount of industry requiring to be supplied with power in the locality.

(4) Where, as at Bellegarde and Fribourg, the power station is 150 or more feet below the factories driven, the telodynamic system has an advantage over some systems, such as the hydraulic system, in that there is no loss of efficiency due to difference of level.

On the other hand, it may be admitted that telodynamic transmissions, simple as they are mechanically, involve considerable cost. The pulley piers require to be lofty and strongly built. The maximum length of span hitherto accomplished is 630 feet, at Bellegarde. Experience has also shown that the cost of maintenance is considerable. The cables must be replaced annually, and experienced workmen are necessary to make the long splices in the ropes.

One distinct disadvantage of the telodynamic system is that no means has been found of directly measuring. by numerous or continuous observations, the amount of power delivered to each consumer. So long as the power is distributed to very few consumers it is possible to assess with practical fairness the charge to each without such measurements of the power. But in proportion as the consumers are more numerous, the defect of the system in this respect becomes more serious.

It is in some cases at any rate a defect of the cable system that the amount of power which it is practically possible to transmit by a single cable is limited. It is not possible by increasing the size of the cable to transmit an indefinitely large amount of power. The cables become too heavy to be manageable, and the pulleys too large in diameter. In the report of the experts advising the town of Geneva in 1889 the limit for one cable was placed at 100 h.p. Messrs. Rieter Brothers place it at 300 h.p., and that amount has in fact been transmitted; no doubt the proper limit varies in different cases.

It is also an inherent characteristic of the cable system that the efficiency decreases rapidly when the distance increases beyond certain moderate limits. On the most favourable interpretation of the experiments the efficiency may be ·96 for 100 yards, or ·93 for 500 yards: efficiencies remarkably high. But the efficiency falls to 0·60 for 5,000 yards: an efficiency by no means remarkably good.

CHAPTER VIII

HYDRAULIC TRANSMISSION

THE distribution of power by pressure water was perhaps first suggested by Bramah, but the origination of a complete system of this kind is due to Lord Armstrong. Although the development of the system has been very gradual, and in spite of the fact that the earliest hydraulic transmissions were of a very limited and local character, it appears that Lord Armstrong from the first contemplated a distribution of power by means of pressure water, in town areas, to many consumers. The supply of motive-power water may be combined with the supply of water to towns for other purposes, the motors being driven by pressure water from the ordinary town mains. It was to such a system that Lord Armstrong first directed attention, and it has the attractive feature that no special network of mains is required for the power water. On the other hand, the pressure is limited to that suitable for ordinary town water supply, and therefore cannot generally exceed 150 to 200 feet of head. Experience has shown that it is better in many cases to have a special system of mains for the supply of power water, and that it is convenient and economical in that case to use a much higher pressure than would be suitable for ordinary town mains. The small mains required for a power distribution can be made to carry safely a pressure impossible in the large mains of an ordinary town supply. For a long time the systems of hydraulic distribution which were constructed were of a local and limited character, and were high pressure systems of this kind. Only in a few towns, pressure water for a small number of motors was obtained from the ordinary mains. In these exceptional cases, the price charged for the water was generally so great that it would have been preferable to use steam or gas engines.

Reverting to high pressure systems, it was soon discovered

that hydraulic transmission had great advantages for driving
lifts, cranes, capstans, dock gates, and similar machines which
work only for short periods and intermittently. For this
particular purpose, it is convenient to use exceptionally high
pressure, with small mains and comparatively small motor
cylinders. Such high pressure hydraulic transmissions were
first erected in connection with docks and arsenals. It was
only after many years that similar systems came to be applied
for power supply over extensive town districts. The conditions
under which high pressure systems first achieved success gave
them a special character which imposes definite limitations on
their application. The high pressure system is almost exclusively an English system, and almost exclusively suitable for
working intermittent machines. For ordinary power purposes
it is less well adapted. Comparatively recently systems of
hydraulic transmission at more moderate pressure have been
carried out, which are better suited to distribute power for
ordinary industrial purposes.

In driving cranes and other intermittently working machines,
the fluctuation in the demand on the mains for pressure water
is very great. Hence, in developing his system, Lord Armstrong
was led very soon to consider the question of the storage of
energy. Reservoir storage for systems in which the pressure is
very great is not generally possible, because no site sufficiently
elevated can be found for the reservoir. Air vessels were considered, but the amount of energy which can be stored in that
way is not very great, and there are practical difficulties. The
pressure in the air vessel varies with the quantity of water in
store, and an air pump must be used to replace the air, which is
absorbed quickly at high pressures, and to maintain the air
cushion. The invention of the hydraulic accumulator met the
difficulty. The accumulator perfectly answers the purpose of
storing such a supply of water under pressure, as is required to
meet the momentary fluctuations of demand on pumping
machinery, which is driving intermittently used motors.

In an article in the 'Mechanic's Magazine,' in 1840,[1] very
interesting now if its date is considered, Lord Armstrong
pointed out that when water is lifted by a pumping engine, it
becomes the recipient of the energy expended in raising it. If

---

[1] See the *Proc. Inst. of Civil Engineers*, vol. l. p. 66.

the same water is used to actuate motors, it renders back the power conferred on it, in its descent to its original level, and thus becomes a medium through which the power of the pumping engine may be transmitted to a distance, and distributed in large or small quantities as required.   Lord Armstrong showed that a continuously working steam pumping engine of comparatively small size was capable of doing a large amount of distributed intermittent work, and he argued that this would be more economical than the employment of a number of steam motors to drive each separate machine.

Soon after this Lord Armstrong invented an hydraulic crane of a type used ever since.   The pressure water acted on a piston, the motion of which was multiplied by reduplicating a chain over pulleys.   In 1845, a crane worked by pressure water from the town mains was erected in Newcastle, and in 1848 similar cranes were used by the North Eastern Railway at their Goods Station in Newcastle.   In 1851, hydraulic transmission was adopted for driving cranes and working dock gates at Great Grimsby, at New Holland on the Humber, and by Brunel on the Great Western Railway.

High and Low Pressure Systems.—Systems of hydraulic transmission are of two distinct types.   (1) There are systems, which for convenience may be termed *low pressure systems*, with reservoir storage.   In these the working pressure is fixed by local conditions, especially by conditions determining the site for the reservoir.   Generally the pressure is not more than 400 to 600 feet.   It is the reservoir storage in these systems which more than anything else makes them suitable for the supply of power for all ordinary industrial purposes, for driving factories or electric light stations, for instance, involving a large continuous demand for power, extending over considerable periods of time.   (2) There are systems, which for convenience may be termed *high pressure systems*, with accumulator storage.   The pressure in these systems is usually 700 to 800 lbs. per sq. in., or 1,600 to 1,800 feet of head.   These systems, in which the reserve of energy is limited in amount, are most suitable for working cranes, lifts, hydraulic presses, and similar intermittently working machines.

Amount of Energy transmitted in Pipes by Pressure Water.—The velocity of water in very long pipes cannot be made great without excessive frictional loss or without incurring danger from hydraulic shock. A velocity of 3 feet per second is very commonly permitted, and perhaps this might be doubled without excessive loss or risk.

Let D be the internal diameter of the pipe in inches; p the working pressure in lbs. per square inch; H the head, in feet, due to the pressure, so that $p = 0.433$ H; r the velocity in feet per second. Then the gross work transmitted is

$$U = \frac{\pi}{4} D^2 p v \text{ foot lbs. per second}$$

$$= 0.34 D^2 H v \text{ foot lbs. per second}$$

or in horses-power

$$\text{h.p.} = .001428 D^2 p v$$

$$= .000618 D^2 H v.$$

GROSS H.P. TRANSMITTED BY DIFFERENT MAINS

Low Pressure System Head, 500 ft.		High Pressure System Pressure, 750 lbs. per sq. in.	
Diameter of main in ins.	Gross h.p. transmitted	Diameter of main in ins.	Gross h.p. transmitted
9	75	3	29
12	133	6	116
18	300	9	260
24	533	12	463

This table is for a velocity of 3 feet per second—the velocity which has been ordinarily permitted. At 6 feet per second the power transmitted would be doubled. The effective power realised in fully loaded motors will be about three-quarters of the amount given in the table.

At 3 feet per second and with a pressure of 500 feet, as at Zürich, a 12-inch main would transmit 133 h.p. and a 24-inch main 533 h.p. Mains of this size can be used with such a pressure. With the high pressure of 750 lbs. per square inch, and at the same velocity as in the case of the London Hydraulic Power System, a 6-inch main transmits 116 h.p., and a 12-inch main, if it could be safely used, would transmit 463 h.p. The horse-power is gross horse-power, without allowing for loss in the motors.[1]

[1] The largest main hitherto used on a high pressure system is $7\frac{1}{2}$ in. internal diameter. But see note, p. 135.

In neither the high pressure nor the low pressure system is the amount of power which can be transmitted by a single main very great. This involves a definite limitation of hydraulic systems. They are best adapted for driving machines working only a fraction of the twenty-four hours, or for motors for small industries not requiring a great amount of power.

Loss of Pressure due to Friction in the Mains.—At a velocity of 3 feet per second the loss of pressure per mile of main due to friction is about 18 lbs. per square inch in a 6-inch main; about 9 lbs. per square inch in a 12-inch main, and about $4\frac{1}{2}$ lbs. per square inch in a 24-inch main. These losses are insignificant on a high pressure system, and not very important on a low pressure system with distances of transmission such as are practically attempted. The losses of energy due to distribution in an hydraulic system, apart from those due to the pumping or motor machines, are so small in most cases that they may be dismissed from consideration without any very serious error.

The loss of pressure measured in feet of head, in a main of length l and diameter D, in feet, at a velocity v feet per second, is given by the equation

$$ h = k \frac{l}{\text{D}} \cdot \frac{v^2}{2g} $$

where k has the following values for clean and corroded pipes : [1]

	New and clean		Old and incrusted
D = 0·5	k = 0·024	to	0·048
1·0	0·022		0·044
2·0	0·020		0·040

The following table gives the loss due to friction, at a velocity of 3 feet per second, per mile of main :—

Diameter of main in ins.	Loss due to friction per mile			
	In ft. of head		In lbs. per sq. in.	
	Clean	Incrusted	Clean	Incrusted
6	35·5	70·9	15·37	30·70
12	16·3	32·5	7·06	14·07
24	7·4	14·8	3·20	6·41

[1] *Machine Design,* Unwin, Part ii. p. 7.

For rough calculations, the loss of pressure per mile may be taken at $107/d$ lbs. per square inch, for pipes in good order. The percentage loss per mile reckoned on the working pressure is as follows :—

NEW AND CLEAN PIPES

Diameter in ins.	Loss per mile in per cent. of total head				
	For pressures in ft. of				
	100	250	500	1,000	1,600
6	35·5	14·2	7·1	3·5	2·2
12	16·3	6·6	3·3	1·6	1·0
24	7·4	2·8	1·4	0·7	0·5

With incrusted pipes the percentage loss is twice as great. The loss in any case is not very important for working pressures of more than 500 feet of head and distances of transmission likely to be attempted. For small working pressures or greater velocity in the main the frictional losses become much more important. That is one reason why high-working pressures are advantageous in hydraulic systems.

Considerations arising out of the Strength of the Pipes.—In all hydraulic systems at high pressure in this country cast-iron pipes have been used, with a peculiar flanged joint having two bolts. Mr. Ellington's experience in London shows that a main of this kind can be made absolutely tight and free from leakage. The largest mains used are $7\frac{1}{2}$ inches in diameter. The working stress in the metal due to the water pressure is 2,800 lbs. per square inch. The mains are usually tested to a water pressure of 2,500 lbs. per square inch before laying, and to a pressure of 800 to 1,600 lbs. per square inch after laying.

Fig. 36 shows at A the form of joint used by Lord Armstrong, and at B a modification introduced by Mr. Ellington. The joint is made tight by a gutta-percha ring. The flanges are placed in a horizontal position in laying. Mr. Ellington found that fractures in the pipes occurred by the breaking off of one of the lugs for the bolts. By placing the lugs a little farther back on the pipe the strength was found to be greater. Probably the slight flexibility of the pipe line this form of joint allows is an important element in its success.

The thickness of the pipes of D inches diameter, for a working pressure of p lbs. per sq. in., is given by the rule—

$$t = 0.000,178 \ \text{D} \ p + \tfrac{1}{4}$$

Thus for $p = 750$ lbs. per sq. in.

D =	4	6	$7\frac{1}{2}$	9	12
$t =$		·78	1·05	1·25	1·45	1·85

The bolts have a stress of about 8,000 lbs. per sq. in. on the net section at the bottom of the thread. If d is the gross diameter of bolt, then approximately, for two bolts

$$d = \frac{\text{D}\sqrt{p}}{11 \cdot 4}$$

Thus for $p = 750$

D = 4	6	$7\frac{1}{2}$	9	12
$d = 1$	$1\frac{1}{2}$	$1\frac{7}{8}$	$2\frac{1}{8}$	$2\frac{7}{8}$

The proportional numbers in fig. 36 are for a unit $= t$.

Probably solid drawn steel pipes could now be used, if a suitable joint for them could be devised.[1] Such pipes were proposed to be used in a project submitted to the Niagara Commission by MM. Vigreux and Feray. For steel pipes a stress

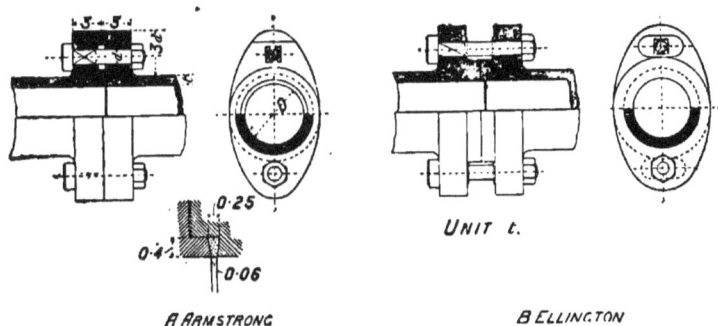

A ARMSTRONG   B ELLINGTON

UNIT t.

FIG. 36.

of 15,000 lbs. per sq. in. might be allowed, and the use of such pipes would much extend the capability of the high pressure hydraulic system. On low pressure hydraulic systems ordinary socket pipes can be used.

---

[1] Mannesman steel tubes of from 5 to 12 inches in diameter are being used to convey water under considerable hydraulic pressure (750 lbs. per sq. in.) at Antwerp. They have stamped steel flanges.

Considerations arising out of the Weight and Cost of the Distributing Mains.—For pipes of equal strength and at a given limiting velocity of flow the weight of mains is simply proportional to the horse-power transmitted. Hence, so far as cost of mains is concerned, the low pressure system is as economical as the high pressure system. Probably, however, if all practical exigencies are taken into account, the cost of mains is somewhat greater for low pressure than for high pressure systems.

Considerations arising out of the Type of Motors driven by the Pressure Water.—On high pressure systems the motors used are almost exclusively pressure engines, that is, motors with reciprocating plungers or pistons. Such motors become extravagantly costly for low working pressures. The greater the working pressure the more conveniently and cheaply is the power produced by motors of this class. Hence the general adoption of pressures of 700 to 800 lbs. per sq. in. in high pressure systems. The pressure engine type of motor is extremely convenient for lifting machinery and hydraulic presses, and even for rotative motors of small size. It is not nearly so convenient when a large amount of power is to be developed continuously for driving a factory. For that purpose turbine motors are much better, being cheaper and more easily regulated. It is true that some of the newer types of turbine, such as impulse turbines and Pelton wheels, can be used even at pressures of 800 lbs. per sq. in. But, on the whole, low pressure reservoir systems are better suited to cases where power has to be developed by turbines.

The Efficiency of Hydraulic Transmission.—Very careful experiments on the efficiency of hydraulic transmission were made at the Marseilles Docks.[1]

The dimensions of the engine and accumulator were as follows:—Engine. Two cylinders 21 ins. diameter, 38 ins. stroke. Accumulators. Ram 17 ins. diameter; pressure $51\frac{1}{2}$ to $52\frac{1}{2}$ atmospheres. The pumps were differential and double-acting. The following table gives some of the principal results of the trials of the engine pumps and accumulator. They have been reduced from metric to English units.

---

[1] 'Docks and Warehouses at Marseilles,' Thomas Hawthorn; *Proc. Inst. Civil Engineers,* vol. xxiv. p. 144.

TRIALS OF ENGINE PUMP AND ACCUMULATOR AT MARSEILLES

Stroke of accumulator in 20 revs. of engine	Time of 20 revs. in seconds	Volume of water forced into accumulator in c. ft. per sec.	Useful work stored in accumulator in ft. lbs. per sec.	Rate of working in h.p.	Slip of pumps per cent.	Indicated h.p. of engine	Ratio of work stored to indicated work of engine
12·40	190	·1030	11,435	21	6·31	26·25	80
12 52	109	·1812	20,120	37	5·38	46·5	79
12·50	90	·2192	24,350	44·8	5 51	58·5	76·5
12·40	90	·2175	24,150	44·5	6·25	58·5	76
12·53	85	·2316	25,850	47·6	5·26	63	75·5
12·52	85	·2324	25,810	47·5	5·38	63	75·4
12·47	70	·2811	31,220	57·5	5·76	77·5	74·2
12·62	66	·3018	33,520	61·7	4·59	83·5	73·9
12·70	60	·3349	37,090	68·3	4·02	92	74·2
12·72	48	·4180	46,420	85·5	3·90	116·5	73·4
12·53	42	·4710	52,300	96·4	5·26	133	72·4
12·75	37	·5437	60,320	111·3	3·65	156	71·3

It appears therefore that at slow speeds of working 20 per cent., and at fast speeds nearly 30 per cent., of the indicated power is lost in engine and pump friction and the friction of the ram packing.[1]

The efficiency of the lifting machinery was also tested. In one case the water used in $1\frac{1}{2}$ hour by five hoists, each loaded with $1\frac{1}{2}$ tons, working simultaneously, was measured, and the engine was indicated at the same time. With 116 i.h.p. of the engine, the energy in the pressure water used amounted to 78·5 h.p. and the useful work done in lifting to 34·6 h.p. Hence the useful work was 20·8 per cent. of the indicated work of the engine, or 44 per cent. of the energy supplied by the pressure water. In another case one hoist was used and the load varied. In this case the ratio of the useful work to the energy supplied by the pressure water varied from 15 per cent. with 1,100 lbs. lifted to 60 per cent. with 4,500 lbs. lifted. The cradle weighed 1,650 lbs. in addition.

Mr. Ellington stated, in his paper on the London Hydraulic Supply System, that 'the practical efficiency (brake h.p. of hydraulic motors) of the hydraulic system may be fixed at from 50 to 60 per cent. of the power developed (indicated h.p.) at the

[1] In the trial of the engine and pumps at Falcon Wharf, given in Mr. Ellington's paper, the i.h.p. was 178·5, and the pump h.p. (allowing 5 per cent. slip) was 139, so that the mechanical efficiency of engines and pumps was 78 per cent.

central station.'[1]   Great weight must be given to anything Mr.
Ellington says in relation to hydraulic power supply, but it must
be pointed out that this assumes a very high efficiency in the
motors.   If the efficiency of engine, pump and accumulator is
taken at 80 per cent., the highest efficiency observed at
Marseilles, and the pressure water is used to drive a continuously
working, fully-loaded turbine, with an efficiency of 75 per cent.,
the resultant efficiency exclusive of any losses in the mains, which
in fact are small, would be 60 per cent.   But with the actual

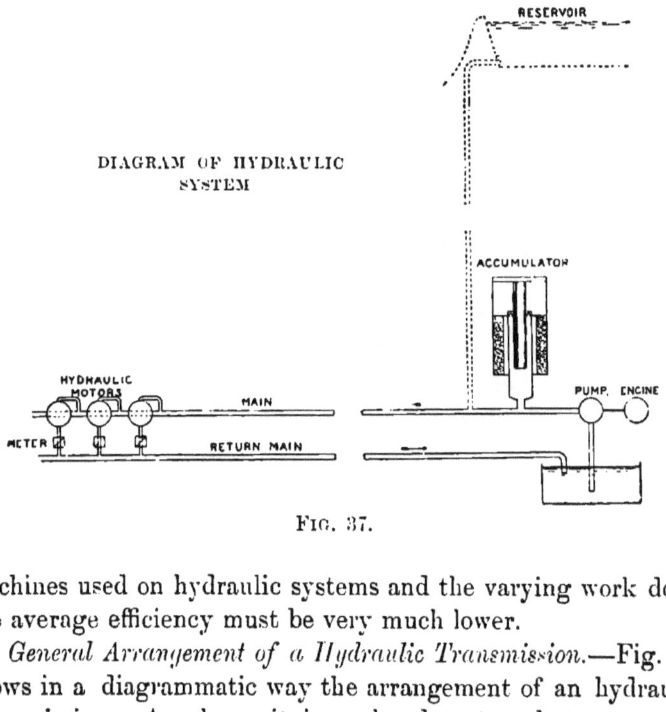

FIG. 37.

machines used on hydraulic systems and the varying work done
the average efficiency must be very much lower.

   General Arrangement of a Hydraulic Transmission.—Fig. 37
shows in a diagrammatic way the arrangement of an hydraulic
transmission.   As shown it is a closed system, but except for
short transmissions the return main is generally omitted, and
the motors discharge water to waste.   Both reservoir and
accumulator storage are indicated.   In extensive systems more
than one accumulator are required, and in that case they are best
distributed throughout the district supplied.

[1] 'Hydraulic Power in London,' *Proc. Inst. Civil Engineers*, vol. xciv.

HIGH PRESSURE SYSTEMS

The highest water pressure usually adopted is 800 lbs. per square inch, but a pressure nearly double this is to be used in Manchester. The distributing mains must in general form a network, so that in case of accident any portion can be shut off by stop valves without affecting the working of the rest of the system. The water used must be pure and free from silt, otherwise the valves and valve seatings and the working parts of the motors are injured. Sometimes the pressure water is obtained from the town mains. At other times water from an impure source is used, but it is filtered before it is pumped into the mains. To obviate injury from frost the pipes must be placed deep enough (three to four feet) underground. In some cases the pressure water is warmed in winter by taking the delivery pipe through the hot well of the engine, or by injecting steam into the suction well. To obviate injury from hydraulic shock, spring-loaded safety valves are placed on the main. Back-pressure valves are also used to prevent a sudden relief of pressure if a main bursts. These are sometimes ball valves hanging in a chamber, which swing so as to close the main if the velocity increases above a safe limit.

Piston Motors.—For ordinary double-acting piston motors let d be the diameter of cylinder and s the length of stroke in inches; let p be the available water pressure in lbs. per square inch; n the revolutions per minute; H the h.p.; η the efficiency. Then

$r = s\,n/360$ is the mean piston speed in feet per second, and

$$\frac{\pi}{4} d^2 p\, r\, \eta = 550\, H$$

$$d = 26 \cdot 46 \sqrt{\frac{H}{\eta\, v\, p}} \text{ inches}$$

η may be taken at 0·7 to 0·8; v is usually about 2 feet per second; the ratio s/d is usually $1\frac{1}{4}$ to $1\frac{1}{2}$.

The Hull Hydraulic Power System.[1]—This was the first scheme for distributing power hydraulically to many consumers. The principal main is 6 inches in diameter and 1,485 yards in length.

[1] See Robinson, *Proc. Inst. Civil Engineers,* vol. xlix.

The joints are flanged joints, with a gutta-percha ring of the kind generally used since. The pumping station is arranged for four 60 h.p. engines, of which two have been erected. Each engine delivers 130 gallons per minute at 700 lbs. per square inch pressure, corresponding to 63·6 effective h.p. There is an accumulator 18 inches in diameter and 20 feet stroke, loaded to 610 lbs. per square inch. The charges were originally intended to be 52*l.* for one crane per annum, and less for several cranes in one warehouse. The charges are, however. by quantity of water supplied as measured by meter. The minimum charge is 8*l.* per machine per annum. The charge by quantity of water used ranges from 8*l.* per annum for 16,000 gallons or less to 200*l.* for 1,200,000 gallons, with special rates for greater quantities. The following short table will give an idea of the way the charges are graduated :—

Consumption of water per quarter	Charge per quarter		Charge per 1,000 gallons supplied
	£	s.	s.
4,000 gallons or under . . .	2	0	10·0
9,000 to 10,000 . . .	3	10	7·0
49,000 to 50,000 . . .	12	10	5·0
99,000 to 100,000 . . .	20	0	4·0
199,000 to 200,000 . . .	35	0	3·5
299,000 to 300,000 . . .	50	0	3·3

The charge for water taken in excess of 300,000 gallons per quarter is 2·5 shillings per 1,000 gallons. For 500,000 gallons per quarter 2·5 shillings per 1,000 is charged for the whole supply.

The London Hydraulic Power Company.[1]—An Act was obtained in 1871 for supplying hydraulic power in London. The rights conferred by the Act remained dormant until resuscitated by Mr. Ellington in 1882. The present company was constituted in 1884. In 1887, twenty-five miles of pressure main had been laid in London streets, and at the present time there are nearly sixty miles of pressure mains. These extend from the West India Docks and Wapping on the east to Kensington on the west; from Mint Street south of the river to Clerkenwell and Old Street on the north.

---

[1] The account of the London Hydraulic Power System is derived partly from papers by Mr. E. B. Ellington (*Proc. Inst. C. E.*, vol. xciv. and vol. cxv.), partly from the reports of the Company.

There are three principal pumping stations: one at Falcon Wharf, a short distance east of Blackfriars Bridge (800 i.h.p.); another at Millbank, Westminster (600 i.h.p.); and one at Wapping (1,200 i.h.p.). A fourth station in the City Road is in course of erection (1,200 i.h.p.). The aggregate power of all the stations, when complete, will be 3,800 i.h.p., of which one-third is reckoned as reserve power in case of repair or accident.

At Falcon Wharf and Millbank all the water is taken from the river, but it is filtered before it is pumped into the mains. At Wapping part is taken from the London Dock, part from a well. After use by consumers it flows into the sewers. The power is available for use night and day all the year round. It is largely used for lifting machinery and for presses and pumps. The company claim that it can be used for electric lighting of particular establishments and for extinguishing fires. For this last purpose Mr. Greathead's injector hydrant or hydraulic intensifier is applied; a small jet of water from the high pressure mains is made to intensify the pressure of a larger jet drawn from the ordinary town mains. A fire stream is so obtained capable of reaching the top of high buildings without employing a fire-engine. In 1892, there were 1,696 machines worked by pressure water from the company's high pressure mains, consuming 6,000,000 gallons per week. The quantity of water used by each consumer is measured by a meter on the exhaust pipe of the machines driven. Parkinson's meter is most used. Siemens' turbine meter is used to some extent, but it is in-accurate under the sudden fluctuations of discharge which occur. Kent's positive meter is also used.

At Falcon Wharf there are four sets of compound pumping engines capable of indicating 200 i.h.p. They are vertical, with one high and two low pressure cylinders, and a pump plunger directly connected to each piston. At 200 feet of piston speed per minute each set of engines will deliver 240 gallons per minute at 750 lbs. per square inch pressure into the accumulator. This corresponds to 120 effective h.p. A nine hours' trial of one set of engines was made in 1887, the engine running at constant speed and the coal used being sea-borne small coal. The boilers are provided with an economiser.

TRIAL OF HYDRAULIC PUMPING ENGINES

Total indicated horse-power	178·5
Piston speed, ft. per min.	221·4
Steam pressure, lbs. per sq. in.	82·5
Evaporation (from and at 212°) per pound of fuel, lbs. .	10·59
Feed water per i.h.p. hour, lbs.	19·79
Coal per i.h.p. hour, lbs.	2·19
Accumulator pressure, lbs. per sq. in.	750
Effective h.p., calculated from water pumped, allowing 5 per cent. slip	139
Mechanical efficiency of engine	0·78
Water pumped per min., gallons	265·7

The engines consume in ordinary work 2·93 lbs. of coal per
i.h.p. hour, which is greater than the result given above in
consequence of the fluctuations of speed.

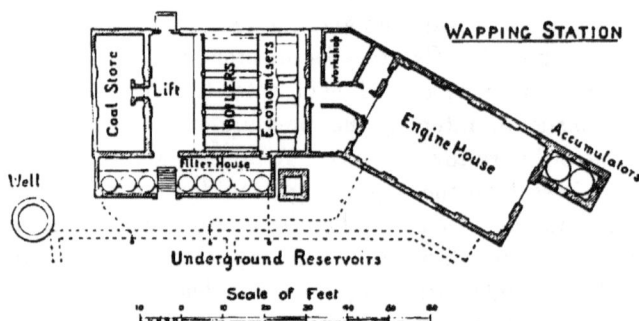

FIG. 38.

There are two accumulators at Falcon Wharf with rams 20
inches in diameter and 23 feet stroke. Each accumulator has
a capacity of storage equal to 2·4 h.p. hours. The filters are
Perrett filters constructed by the Pulsometer Company. The
filtering material is compressed sponge. The sponge is kept
under a pressure of about 4 lbs. per square inch by a special
hydraulic ram. It is cleansed every four to six hours by reversing
the direction of flow and by alternately compressing and
releasing the pressure on the sponge.

The pumping station at Wapping,[1] of which a plan is shown
in fig. 38, is a more recently constructed and larger station than

[1] The Wapping Station was described in the *Engineer* for January 20,
1893.

that at Falcon Wharf. The water pumped is obtained partly from a well sunk into a gravel bed, partly from the London Dock. The pumping from the well into a tank over the boiler-house is effected by low lift pumps worked hydraulically by the pressure water. From this tank it passes through 'Torrent' filters constructed by the Pulsometer Company to underground reservoirs. From this it is lifted by the condenser circulating pumps to another tank above the boiler-house, whence it is pumped into the mains. The reservoir capacity is 800,000 gallons. The engine-house contains six sets of vertical inverted triple expansion engines with cylinders 15 inches, 22 inches, and 36 inches in diameter and 24 inches stroke. Each piston drives a single-acting plunger pump with ram 5 inches in diameter direct from the crosshead. The working steam pressure is 150 lbs. per square inch, and the hydraulic pressure 800 lbs. per square inch. Each set of engines will deliver 300 gallons of water per minute at a piston speed of 250 feet per minute. All the cylinders are jacketed.

In a test trial the engines are stated to have worked with 14·1 lbs. of steam and 1·27 lb. of Welsh coal per i.h.p. hour. The water passes from the pumps to two accumulators, with rams 20 inches in diameter and 23 feet stroke. One of the accumulators is loaded to a slightly heavier pressure than the other, so that one accumulator rises a little in advance of the other. The more heavily loaded accumulator automatically shuts off steam when at the top of its stroke.

Charges for Pressure Water for Power Purposes.—The London Hydraulic Company make a minimum charge of 1*l*. 5*s*. per quarter per machine. For consumers using more than 3,000 gallons per quarter there are graduated charges of which the following short table gives a sample:—

Gallons used per quarter	Charge		Cost of pressure water per 1,000 gallons
	£	s.	s.
3,000 or under .	1	5	8·3
10,000 . . .	3	10	7·0
50,000 . . .	12	10	5·0
100,000 . . .	20	0	4·0
200,000 . . .	31	5	3·1
300,000 . . .	42	10	2·8

An excess over 300,000 gallons per quarter is charged at 2s. per 1,000 gallons. Consumers using more than 500,000 gallons per quarter are charged 2s. per 1,000 gallons all round. Rates are further reduced for still larger quantities, and the minimum rate is 1·5s. per 1,000 gallons.

With these charges the cost of power for the kind of work for which an hydraulic system is best suited is small. Thus it is often less than one farthing per ton lifted 50 feet. On the other hand, it is necessary for the purpose of this treatise to consider the cost of power distributed by different methods on some common basis. It is almost unavoidable to take the cost of power exerted continuously through the working day. If the cost of power supplied by the Hydraulic Power Company is reckoned, for machines working 3,000 hours per year, then the cost is larger than that of power obtained in other ways. The comparison of the cost so reckoned is instructive, although it may be in fairness pointed out that Mr. Ellington in his paper stated that he had never advocated the supply of power by the London Hydraulic System for continuously working engines to any large extent.

To obtain one effective h.p. during 3,000 hours per annum, allowing an efficiency of 80 per cent. in the motor, 437,500 gallons of water are required. Hence a consumer, taking 50,000 gallons per quarter, would get the equivalent annually of 0·457 effective h.p. for 3,000 hours, and would pay for it at the rate of 109l. per h.p. per annum. A consumer taking 300,000 gallons per quarter would get the equivalent of 2·743 effective h.p. for 3,000 hours, and would pay at the rate of 62l. per effective h.p. per annum. A consumer taking 500,000 gallons per quarter would get the equivalent of 4·573 effective h.p. for 3,000 hours, and would pay at the rate of 43l. 15s. per effective h.p. per annum. It must be remembered that this is the cost for pressure water only, and does not include meter rent or interest on the cost of the motors.

There is one other respect in which the statement just made, unfavourable as it is to the use of pressure water as an agent in distributing water power, is nevertheless too favourable. It follows from the incompressibility of water that nothing like expansive working is possible in a water motor. In the case of all reciprocating motors, and these are almost the only motors used with high pressure water, and with the small exception of the special engine of Mr. Rigg mentioned above, it may be said broadly that all the motors on high pressure hydraulic systems use the same quantity of water, whether they are lightly loaded or fully loaded. The consequence is that not only is pressure water expensive as an agent for distributing power, when it is used as economically as is possible in fully loaded motors, but the cost is again increased because in practice most of the motors work usually at less than full load. If we take the average load on the motors to be not more than two-thirds the full load, then the cost of the power is increased 50 per cent.

The Liverpool Hydraulic Supply System.—In Liverpool pressure water from the town mains was used for working hydraulic cranes as early as 1847. From an interesting paper by Mr. Joseph Parry,[1] it appears that the use of hydraulic power in this way made very slow progress. In 1877, the number of hydraulic machines supplied from the town mains was 89. At the present time there are 162 machines worked by water from the town mains, consuming 125,600,000 gallons per annum. Taking the mean pressure at 70 lbs. per sq. in., this is equivalent to 82,710 effective h.p. hours, or to 27 effective h.p. for 3,000 hours in the year—a rather insignificant amount. The average charge for working a goods hoist is 13*l*. per annum, or only 10*d*. per hoist per day, a small cost for the convenience afforded. The charge for water is 7*d*. per 1,000 gallons. At this rate the charge is equivalent to 120*l*. per effective h.p. per year of 3,000 hours. Experiments on the quantity of water used by some hoists showed the cost to amount to from 6*d*. to 10*d*. per ton lifted 50 feet.

There is also in Liverpool a high pressure system, which is to be extended. Experiments with some hoists worked on this system showed the cost to be from 1¾*d*. to 2¼*d*. per ton lifted 50 feet. Mr. Parry comes to the conclusion that hoists worked from the town mains cost more than those on the high pressure system, when the charge for water on the high pressure system does not exceed 5*s*. per 1,000 gallons.

The Birmingham Hydraulic Power System.—In Birmingham, as in Liverpool, water has been supplied from the town mains

[1] 'The Supply of Power by Pressure from the Public Mains,' *Proc. Inst. of Mechanical Engineers.*

.to work lifts. In 1888, there were 61 lifts and hoists thus
worked, using 80,000 gallons per day, and yielding to the Water
Committee of the Corporation about 1,000*l.* a year. Since that
time, a high pressure system has been carried out, which has
the peculiarities that it belongs to the Corporation and that the
pumping is done by gas engines.[1]

At the pumping station there are three sets of triple
hydraulic pumps, working to a pressure of 730 lbs. per sq. in.
These are driven by three 'Otto' gas engines, nominally of 12,
20, and 20 h.p., but capable of developing an aggregate of
about 100 h.p. Ordinary lighting gas is used. The pumps
deliver into two 6-inch mains. There are three hydraulic
pumps, having each three plungers. For one the plungers are
2½ ins. diameter and 9 ins. stroke. For the others the plungers
are 3 ins. diameter and 12 ins. stroke. Each gas engine drives
a counter-shaft by a belt, and this shaft drives the pump crank-
shaft by gearing at 49 revs. per minute. There are two accumulators with 20-inch rams and 20 feet stroke. A small
'Brotherhood' engine, worked by the pressure water, is used in
starting the gas engines.

Manchester Hydraulic Power Supply.—At Manchester a
combined scheme for supplying electricity and high pressure
water is being carried out. A pressure of 1,600 lbs. per sq. in.
in the hydraulic system is to be used. It is hoped that there
will be economy in working the electricity and pressure water
supply from the station.

LOW PRESSURE HYDRAULIC SYSTEMS

The Zürich Works.—The Zürich installation is a complex
and very interesting one.[2] It was the earliest example in
Switzerland of the application of hydraulic power, partly to
pump a supply of potable water, partly to furnish motive power,
from the same central station. It has grown gradually, and of
late has been greatly extended. It comprises machinery driven
by turbines for furnishing (*a*) a water supply to the town of
Zürich; (*b*) a supply of motive power transmitted by wire rope;

[1] See *Engineering*, February 12, 1892.
[2] See Preller on 'The Zürich Water Supply Power and Electric Works,'
Proc. Inst. Civil Engineers, vol. cxi.

(c) a supply of motive power transmitted by comparatively low pressure water from the town mains; (d) a supply of motive power transmitted hydraulically, from a special reservoir at comparatively high pressure; (e) an electric central station, also driven by water power.

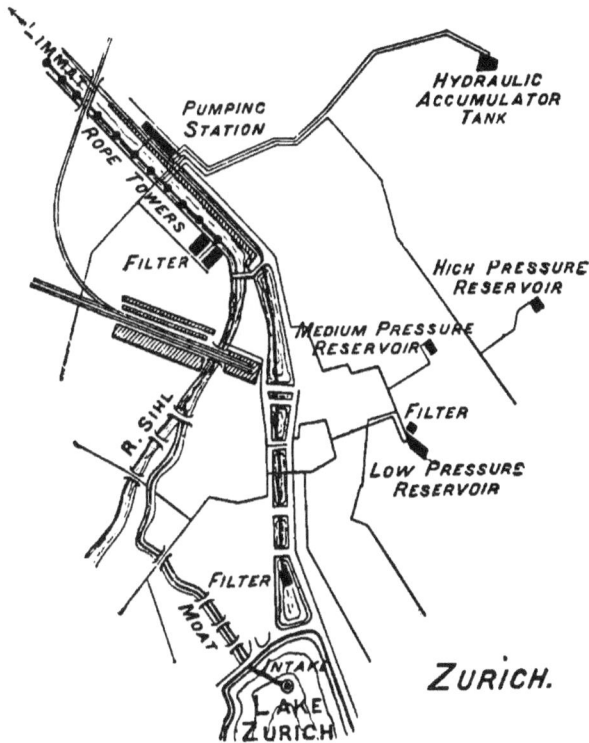

FIG. 39.

When the works (fig. 39) were first established, the water supply of Zürich was obtained from a filter in the bed of the river Limmat near its exit from the lake. This water was pumped by turbines erected a little further down stream. There being surplus water power, a telodynamic transmission was erected, and part of the motive power was distributed to factories along the riverside. In 1884 the quality of the water was found to be inferior. After extensive investigations, it was decided to obtain a new supply of potable water from an intake in the lake, and to use the old water supply for motive power purposes only.

The fall available in the Limmat at the pumping-station, and the available volume of flow, are as follows:—

—	Fall, feet	Volume of water flowing in the river, c. ft. per sec.	Gross water power, h.p.
High-water level in summer	4·92	2,295	1,300
Mean „ „ „ „	8·20	1,660	1,570
Low „ „ „ winter .	10·50	1,059	1,280

The effective power delivered by the turbines in the river is as follows:—

	H.P.
For pumping filtered potable water . . .	237
Supplying motive power by pressure water . .	128
Driving the wire rope transmission	227
Supplying hydraulic motive power for electric lighting station	444
Total	1,036

There are two reserve steam-engines, of 300 i.h.p. each. to provide for a deficiency of water power. At the pumping station there are eight pressure (Jonval) turbines working, up to from 96 to 110 h.p., according to the state of the river. There are also two newer turbines of about 175 h.p. each. The turbines have vertical shafts, and each pair drives by bevil wheels a common horizontal shaft, which runs at 50 revolutions per minute in the case of the earlier turbines, and at 66 revolutions per minute in the case of the two last erected. From these shafts a horizontal main shaft, 328 feet in length, and running at 100 revolutions per minute, is driven. To this main shaft any of the pumps can be coupled. The earlier turbines cost, with gearing, about 12l. per h.p.; the two larger turbines about 7l. per h.p. There are at present in operation seven sets of horizontal double-acting 'Girard' pumps. The total pumping capacity is 8,143,000 gallons per day. The water for driving the turbines is obtained by a weir in the Limmat, which deviates the water into a canal formed by a longitudinal embankment in the river. Sluices divide the head-race channel from a tail-race channel formed in a similar way.

The pumps supply the following reservoirs :—

	Height above pumps ft.	Capacity in c. ft.	
		Present	To be increased to
Low level	154	205,000	—
Intermediate level . . .	300	68,500	141,000
High level	485	10,600	21,190
Reservoir for high pressure power water	528	353,170	529,000

From the town mains water is supplied to work 180 small motors. The total power thus supplied is about 187 h.p., and its cost is about 4·4d. per h.p. hour.

The principal supply of power, apart from that distributed by wire rope, is obtained from pressure water derived from the old Limmat filter-bed, and pumped to the special high level reservoir. This water is pumped chiefly during the night. The reservoir is about 6.000 feet from the pumping station, and is supplied by an 18-inch main. The effective pressure at the motors is about 475 feet, and the distributing mains have an aggregate length of 15,000 feet.

The charge for this pressure water for power purposes varies from 0·6d. per h.p. hour, when at least 50,000 h.p. hours are taken in the year, to 1·25d. per h.p. hour, when less than 20,000 h.p. hours are taken in the year. For 3,000 working hours in the year the charge is from 7l. 10s. to 16l. per h.p. per annum. The water supplied in this way now amounts to 42,380,000 c. ft. per annum, yielding altogether about 900,000 h.p. hours. The total receipts are 1,200l. per annum, or 1·08d. per 1,000 gallons.

Besides this supply of pressure water to various consumers, the Electric Station (fig. 40) is ordinarily to be driven by pressure water from the same high-level reservoir. For this purpose two impulse turbines of 300 h.p. each have been erected for driving dynamos, and two smaller turbines of 30 h.p. driving exciting dynamos. Alternatively, if the supply of pressure water fails, the dynamos can be driven by the river turbines, or by the reserve steam-engines.

The Hydraulic Works and System of Hydraulic Power Supply at Geneva.—There is now in operation at Geneva one of the

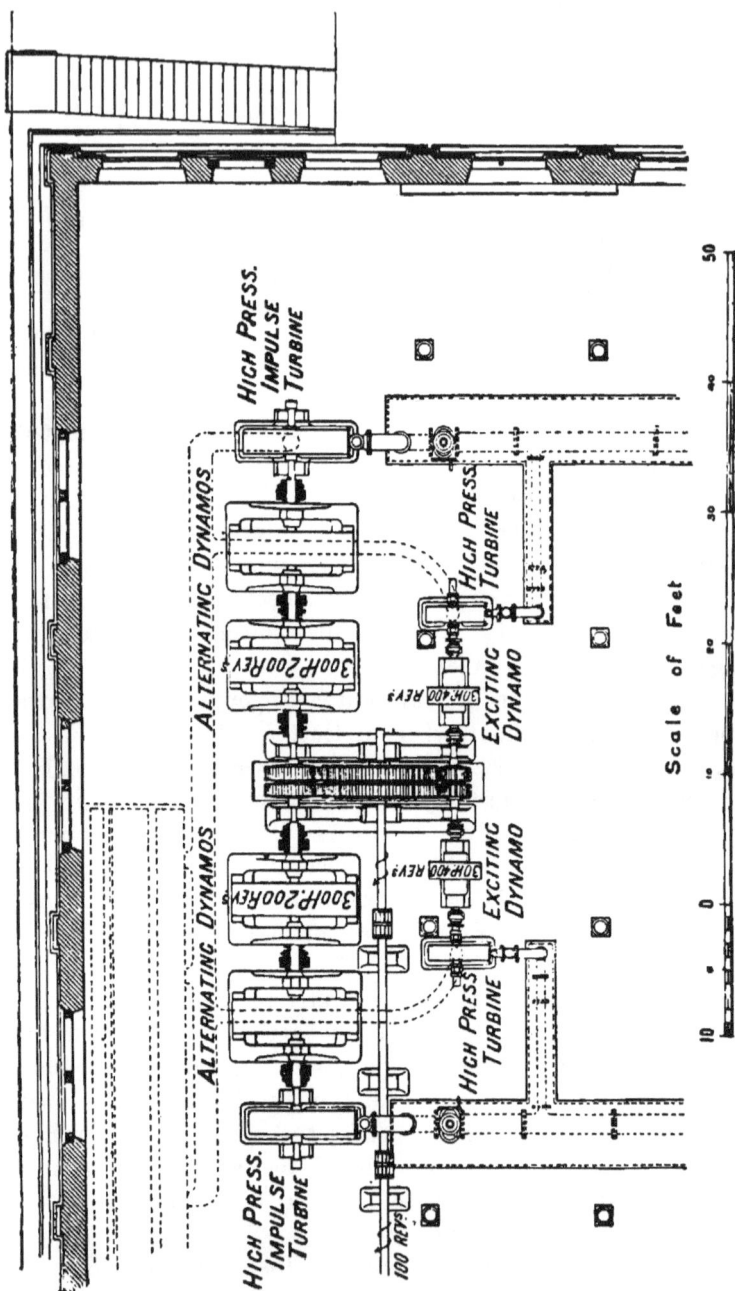

HIGH PRESS. IMPULSE TURBINE

ALTERNATING DYNAMOS

HIGH PRESS. TURBINE

EXCITING DYNAMO

300 H.P. 200 REVS

30 H.P. 400 REVS

ALTERNATING DYNAMOS

300 H.P. 200 REV.S

30 H.P. 400 REV.S

EXCITING DYNAMO

HIGH PRESS. TURBINE

HIGH PRESS. IMPULSE TURBINE

100 REV.S

Scale of Feet

FIG. 40.

most remarkable hydraulic power stations in the world. The water of the river Rhône, near the point where it flows out of Lake Leman, is employed to drive a number of large low pressure turbines, giving a total of 4,500 effective h.p. These turbines pump pure water obtained from the lake into two systems of mains. The older of these, termed the low pressure system, the pressure at the pumps being 160 to 200 feet, is an extension of a previously existing system of mains used for supplying potable water to the town of Geneva. Although some of the water pumped into this system is used for power purposes, it is chiefly intended to supply water for domestic and municipal purposes. The second system of mains, termed the high pressure system, the pressure at the pumps being 460 feet, supplies potable water to some districts not reached by the low pressure system, but it is specially intended to afford a supply of water for motive power purposes to the entire area of the town. The demand for water, both on the low and high pressure systems, is a fluctuating demand, large during the day, and very small during the night. Hence, if the turbines in the Rhône were employed solely in pumping into the mains, they would not be continuously working, and a large part of the water power of the Rhône would be wasted. To meet this difficulty an important storage reservoir has been constructed at Bessinges, about 4 kilometres from Geneva. The turbines pump water up to this reservoir at night, and at times when the demand for power for other purposes is insufficient to keep them fully employed. The energy derived from water flowing back from the Bessinges reservoir through the high pressure system represents part of the water power of the Rhône, which would necessarily have been wasted if this means of storage had not been provided.

The works at Geneva have gradually developed under special local conditions. In spite of natural and political isolation, manufacturing industries have for centuries flourished at Geneva. That they did so is partly owing to the fact that cheap water power could be obtained by simple forms of water-wheel placed in the ample and rapid Rhône, flowing past the town. An industrial quarter gathered along the banks of the river, and factories were built even in the stream itself. As the population increased a water supply was required. The small aqueducts of spring water became insufficient, and further recourse was had to the motive power of the Rhône. From the beginning of the eighteenth century, water-wheels placed in the Rhône pumped a water supply into the town.

Then arose an antagonism to the utilisation of the motive power of the Rhône, which for two centuries hindered the progress of industrial enterprise at Geneva, and threatened at times to destroy the existing industries. The properties of riparian owners on the shores of Lake Leman were from time to time injured by the rising of the lake level. It was not unnatural that the landowners should attribute the disastrous inundations from which they suffered to the obstacles created at the outlet of the lake, that is to the bridges and buildings, and especially the factories and water-wheels in Geneva. Complaints were addressed by the Canton Vaud to the Federal Government at Berne of damage caused by the works at Geneva. Then arose a question of arrangements necessary to regulate the lake level, and to facilitate, in time of flood, the discharge of the water. After 1875, the project of utilising the motive power of the Rhône took a new magnitude and importance, from the combination with it of plans for regulating the level of Lake Leman, and so ending a long and bitter controversy.

Another local circumstance had great influence in determining the ultimate form of the project for the utilisation of the motive power of the Rhône. In 1871, Colonel Turrettini,[1] the engineer under whose direction the present works have been constructed, had applied to the town council of Geneva to place small pressure engines on the mains of the then existing low pressure water supply. The plan of obtaining motive power in this way proved so successful and convenient that, in 1880, there were 111 motors at work, using 34 million cubic feet of water annually, and paying a yearly rental for power water of 2,000l. The cost of the power at that time to consumers was at the rate of 36l. to 48l. per h.p., per year of 3,000 working hours.

In 1878, a private firm asked the concession of a monopoly of the motive power of the Rhône at Geneva, on condition of carrying out works necessary for facilitating the discharge from

---

[1] *Utilisation des Forces motrices du Rhône et Régularisation du Lac Leman*, Th. Turrettini, Ingénieur Conseiller Administratif délégué aux travaux; Genève, 1890. This admirably illustrated memoir fully describes the origin, progress, and details of all the works at Geneva.

the lake and regulating the lake level. A similar offer was made in 1881. But there grew up a feeling that such works should be carried out by, and for the profit of, the town itself. Finally, after many studies, the contract was given by the municipality in 1883 to M. Chappuis to construct under their direction the present works.

These works have cost altogether 283,000l. Of this sum a fraction has been paid by owners of land on the shores of the lake, and part has been expended in constructing new sewers required in consequence of the alterations of river level. Deducting these items, the cost of utilising the motive power of the Rhône has already amounted to nearly 200,000l.

The scheme included the clearing away of all obstacles to the free flow of the river, and the division of the river by a longitudinal embankment into two portions, one forming a head-race to the turbines, fig. 41, the other, which was straightened and deepened, forming an outlet for the surplus water from the lake. Between the two

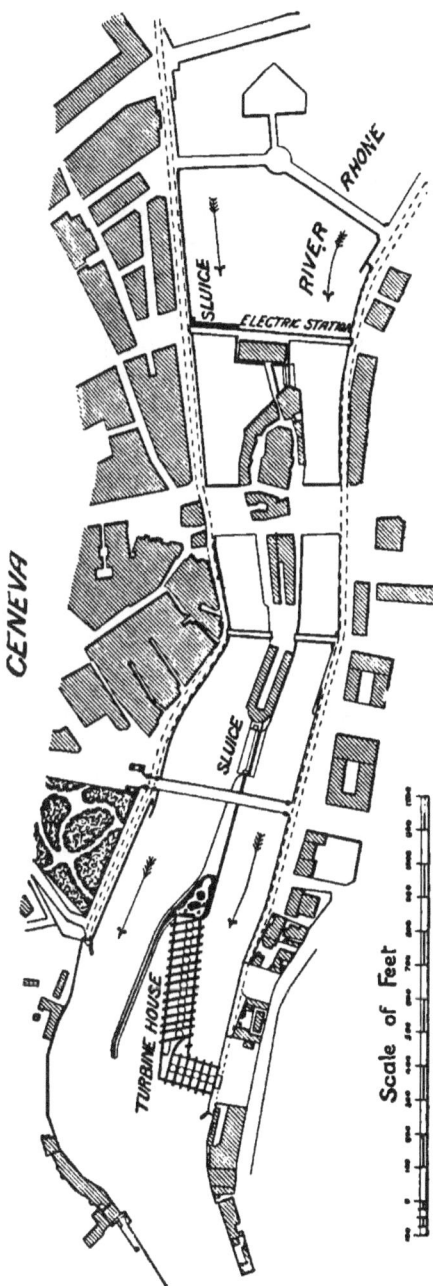

FIG. 41.

divisions of the river bed are movable sluices, which keep up
the water in the head-race channel, or discharge surplus water
into the tail-race channel, according to the condition of the lake.
The scheme also included a complete reconstruction of the old
pumping system for the low pressure water supply ; the creation of the new system of high pressure water supply, and the
provision of motive power by hydraulic transmission to the
industries of the town.

The Low Pressure River Turbines.—The turbine and pump
house is placed at the end (fig. 41) of the left-hand channel
or head-race. The turbines are of 210 h.p. each. and 14
groups of turbines and pumps have been erected. Four more
groups of somewhat greater power are expected to be erected
within the next five years. The turbines are Jonval pressure
turbines, constructed by Messrs. Escher Wyss & Co.. of Zürich.
They have vertical shafts, and each turbine drives from a crank
two horizontal double-acting ' Girard' pumps. placed at right
angles. Fig. 42 shows a cross-section of the turbine and pump
house.

The head at the turbines varies from 5·5 feet, when the river
is in flood, to 12·14 feet when the volume of flow is smallest.
With most forms of turbine this would involve a considerable
variation of the normal speed, or speed of greatest efficiency.
The turbines are skilfully arranged to meet this variation of
head. The turbine-wheel and its corresponding system of
guide-passages are arranged in three concentric rings. When
the fall is great and the quantity of water used is smallest
the outer ring only is open, and the water acts at a large radius.
As the fall diminishes the second ring is opened and the mean
radius at which the water acts is smaller. In the lowest
conditions of the fall, when most water must be used, all three
rings are open, and the mean radius at which the water acts
is smaller still. The number of rotations of the turbine depends directly on the velocity due to the head. and inversely on
the radius at which the water enters. Hence, as the radius
diminishes as the head diminishes, a fairly constant speed of
rotation is obtained. The adjustment is such that with the
highest fall the normal speed is 27 revolutions per minute, and
with the lowest fall 24 revolutions per minute. a variation not
practically serious in working pumps.

The fixed distributor over the turbine-wheel is 13·78 feet
in external diameter and 5·74 feet in internal diameter. It is
divided into three rings having fifty-two guide passages in the
outer ring, forty-eight in the middle ring, and forty in the inner
ring. The external ring has no regulating sluices, regulation
being effected when that ring only is open by the sluices in the
head-race. The other rings are arranged so that over one semicircle the orifices open vertically on an annular plane surface,
and over the other semicircle they open horizontally on a

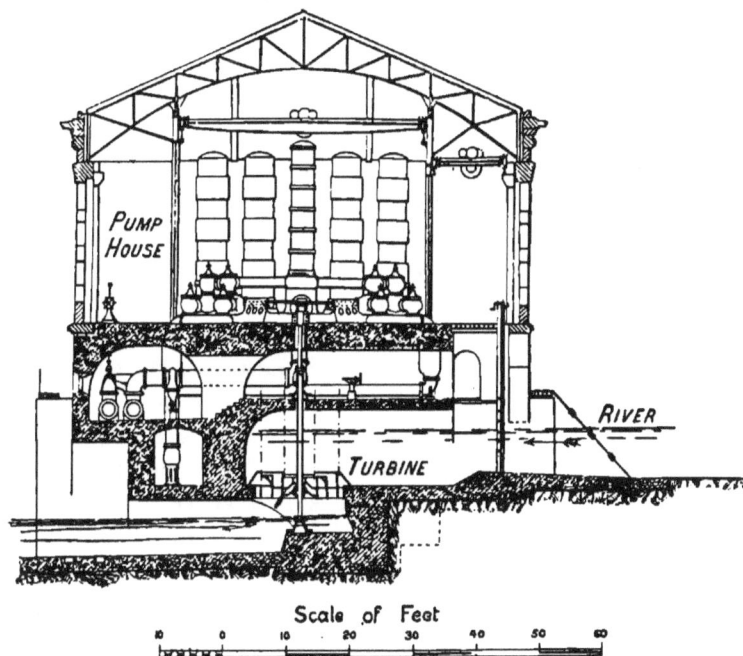

Scale of Feet

FIG. 42.

cylindrical surface. Each ring of passages has two regulating
sluices, one a semicircular annular plate for the orifices opening
vertically, one a semi-cylinder for the openings which are horizontal. Each sluice can be fully opened without interfering
with the openings corresponding to the other. The sluices are
worked by gearing. The turbine-wheel is of cast iron, in two
halves. It has wheel passages corresponding to those in the
distributor. The effective section of flow through each turbine
is :—

Outer ring, fifty-two passages, each 11·02 ins. × 2·95 ins., giving a total of 11·75 sq. ft.

Middle ring, forty-eight passages, each 17·72 × 2·60 ins., giving a total of 15·35 sq. ft.

Inner ring, forty passages, each 17·72 × 2·36 ins., giving a total of 11·63 sq. ft.

In low conditions of the river, when the fall is greatest (12·14 ft.), the turbine must discharge 211·9 c. ft. per second. The relative velocity of discharge is $0·65\sqrt{2g} × 12·14 = 18·17$ ft. per second. Then the outer ring affords sufficient area.

In high conditions of the river, when the fall is 5·51 ft., the turbine must discharge 471·4 c. ft. per second. The relative velocity of discharge is 12·25 ft. per second. Then the three rings give sufficient area.

The vertical support of each turbine consists of a fixed wrought-iron pillar, carrying at its top a steel step for the pivot and a steel revolving hollow shaft hanging from the pivot at the top. The pivot is 6 inches in diameter. A crank at the top of the shaft drives two 'Girard' double-acting pumps placed at right angles, from a single crank-pin. The 'Girard' pump consists virtually of two plunger pumps placed end to end, the advantage being that the stuffing-boxes for the plungers are accessible and there is no internal packing. The two pumps discharge into a single air-vessel placed between them. The diameter of the plungers of the low pressure pumps is 1·41 feet, that of the high pressure pumps 1·08 and 0·85 feet. The stroke is 3·61 feet, and the mean velocity 188 feet per minute. The valves are ring valves with leather faces. The high pressure pumps supply mains of 20-inch diameter in one direction and of 24-inch and 16-inch in the other. The low pressure pumps supply two mains of 20-inch diameter.

The High Pressure Reservoir at Bessinges.—When the high pressure system was first put in operation, a constant pressure was maintained in the mains by constantly pumping in excess of the demand and allowing the surplus to flow away through a relief valve. This involved a constant waste. To further moderate fluctuations of pressure four large air vessels (additional to those at the pumps) were erected. These were 5 feet in diameter and 39 feet high, and were kept charged with air by a 'Colladon' compressor. When it became a question of driving the electric station by turbines driven by water from the high pressure system, the need of a storage reservoir became pressing. At 4 kilometres from Geneva, a site was found at an elevation of 390 feet above the lake, and it was decided to construct a reservoir, capable of storing the discharge of three groups of pumps working through the night. The discharge of the pumps is 34,000 c. ft. per hour or 442,000 c. ft. for thirteen hours, during which, if there were no means of storage, they would be put out of action.

The reservoir is a covered reservoir capable of containing 453,000 c. ft. It stores, therefore, 5,573 gross h.p. hours of energy. Allowing for the loss at the motors driven by the pressure water, the reservoir will furnish about 800 effective h.p. for five hours. It serves as a perfect regulator of pressure. A float with electric signal and recording apparatus shows constantly in the pump-house the condition of the reservoir.

Hydraulic Pressure Relay or Compensating Pressure Regulator. The 16-inch pipe main from the pumping-station to the reservoir at Bessinges being 4 kilometres in length, there would be a difference of pressure in the mains in Geneva equivalent to the friction of 8 kilometres of main, according as water was being pumped up to or flowing back from the

FIG. 43.

reservoir. This would not have been very serious if all the motors driven by the water had been supplied by meter. But the larger motors are supplied by gauging, the quantity of water used being computed from the area of the orifices of discharge. Variations of head would have involved a variation of the quantity of power developed at the motors amounting to 20 per cent. This would have hindered the development of that method of estimating the charge for water.[1]

To prevent this variation of head Colonel Turrettini devised a centrifugal pump relay, shown diagrammatically in fig. 43, and in

---

[1] Any difficulty in gauging the quantity of water used by large consumers could now be obviated by measuring the water directly by the very ingenious and accurate 'Venturi' meter of Mr. Herschel. It was used at Chicago to measure the supply of water from the Waukesha Spring to the Exhibition.

fig. 44, which comes into action automatically and increases the pressure whenever the water is returning from the reservoir to the town. The centrifugal pump, which forms part of the main, is driven by a turbine so regulated for speed that a constant pressure is obtained on the town side of the pump. The pump receives at the maximum 635 c. ft. of water per minute, and can give to the stream passing through it an increase of pressure of 30 feet. The turbine works with 380 feet of head, and can exert 120 h.p.

The sluices of the turbine are governed by an automatic pressure regulator. The pressure in the main acts on a piston controlled by a spring. According to the position of the piston the turbine sluices are opened more or less. The movement of the piston actuates the valve of an hydraulic relay, which operates the turbine sluices. During the filling of the reservoir the centrifugal pump is at rest, the water merely flowing through it. When water flows back from the reservoir, the turbine begins to drive the pump so as to increase the pressure in the main. The arrangement has worked with perfect success. Fig. 44 shows a cross-section and plan of the hydraulic relay and pipes connecting it to the pumping main.

The Motors used in Geneva.—The original motors used in Geneva were ' Schmid ' pressure engines, and these are still used for small powers. They use a quantity of water which depends on the speed only, and not on the work done. Hence they are uneconomical with light loads. They are convenient and cheap, they can be run at any speed, and they act as meters of the quantity of water used. A counter on the pressure engine, recording the number of revolutions, gives the means of ascertaining accurately the quantity of water consumed. At full load their efficiency is 80 per cent.

For all larger motors impulse turbines are used. The maximum efficiency of these is 75 per cent., and it is not much less with light loads. They occupy little space, and can be perfectly governed to constant speed by the ingenious relay governors of Messrs. Faesch & Piccard. In Geneva, the question of speed regulation was found to be an important one. The industries connected with watch-making required motors running at constant speed.

The Electric Lighting Station.—In 1887 the Geneva city council came to an arrangement with a company for supplying electricity.

It was part of the arrangement that the company should use pressure water supplied by the town, as motive power in all its installations. The pressure water is supplied to the company by meter, at a price of 2 centimes per metre cube, with a minimum of 400,000 c. metres annually. This is equivalent to a little more than 3*l.* per effective h.p. per annum. The advantage to the town is that their pumping machinery can be run constantly night and day, the energy which would otherwise be

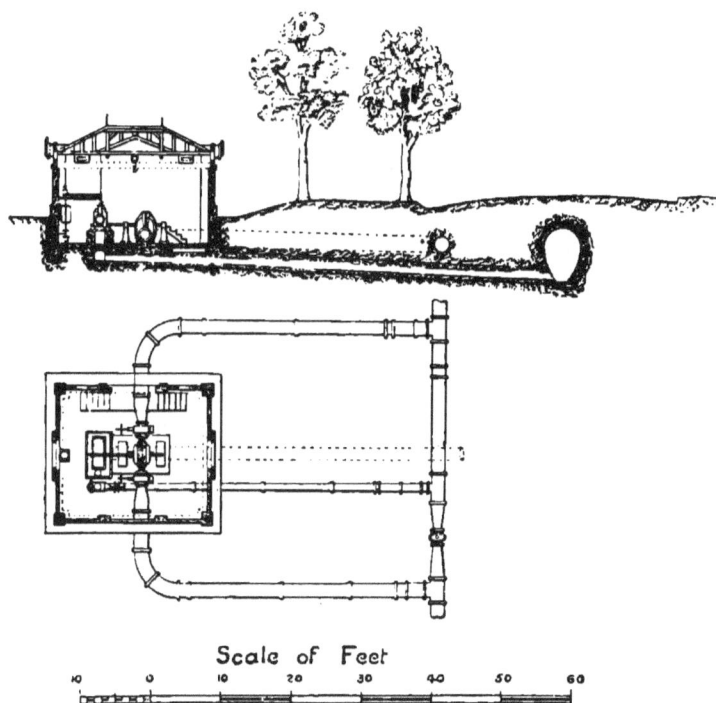

Scale of Feet

FIG. 41.

wasted being stored. The electric company, on the other hand, get power at a cheap rate, and their turbines being driven by high pressure are convenient and cheap, and run at an extremely constant speed.

Under the arrangement an electric station has been installed in the old pumping station, no longer required for its original purpose (see fig. 41). There are three impulse turbines of 200 h.p. each, and each turbine drives two dynamos directly coupled

to it by 'Raffard' couplings (fig. 45).[1] There is also a 25 h.p.
turbine and dynamo for day work. It is the system of reservoir
storage which makes this hydraulic driving of the dynamos
possible and economical. They could not be driven so conveniently by the large low pressure turbines in the river, with
the very varying head which they have to utilise, nor could
power be spared to drive them except by utilising the motive
power of the flow of the river through the night.

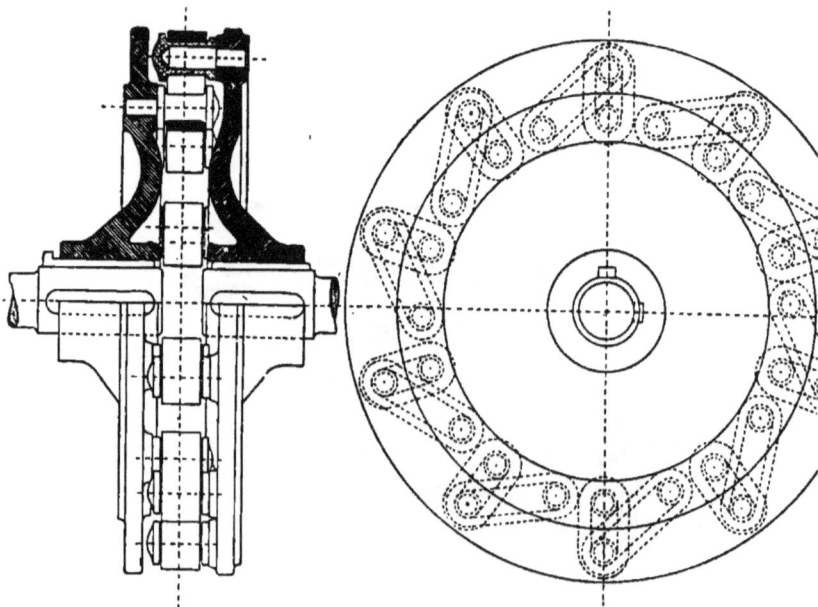

FIG. 45.

*The Installation of the 'Compagnie Hydro-Electrique' at
Antwerp.*—A very remarkable scheme for the hydraulic distribution of power is being carried out at Antwerp, and the author

---

[1] In the 'Raffard' couplings two discs on the shaft are connected by india-
rubber bands. which have a small initial tension when the shafts are not
driving. The figure shows the position of the bands when the shafts are at
rest and the position they take when driving. The coefficient of elasticity of
the rubber may be taken at $E = 119$ lbs. per sq. in. The limiting stress when
driving should not exceed 50 lbs. per sq. in. The coupling has the advantage,
electrically, of being an insulating coupling. It has been used extensively,
even for transmitting large amounts of power, and works very satisfactorily.
It is most suitable for connecting shafts running at least 250 revolutions per
minute.

is indebted to Messrs. Carels Frères, of Ghent, who have constructed the steam-pumping machinery, for the following details. The plans are based on the investigations of the late Professor François von Rysselbergh, whose studies and improvements, especially in electrical science, are well known.

Von Rysselbergh recognised the inconveniences of the distribution of electricity from a single station over a wide area. With low tension continuous current, the cost of the distributing mains is enormous : and with high tension, alternating currents, and transformers, he believed the dangers to be serious. He sought, therefore, for means of distributing electricity at moderate tension without incurring too great a cost in the network of mains. He was led to a system which may be briefly described as follows :—

(1) Hydraulic transmission is adopted as the primary means of distributing motive power over the district. Water under a pressure of 770 lbs. per sq. in. is pumped into the distributing mains by steam engines.

(2) The pressure water is used to actuate motors which generate motive power for industrial purposes, or which drive dynamos producing electricity in electric sub-stations scattered at convenient positions over the district.

(3) The hydraulic motors are turbines of a new and special type, of small dimensions, not costly, and easily managed. It is stated that the 'von Rysselbergh' motor will drive a dynamo, so that the difference of potential at any point of one of the electric circuits never varies more than 2 volts from the mean pressure of 110 volts.

The system, therefore, serves a double purpose. It affords a means of distributing motive power in a form very convenient for small industries. It also permits the convenient distribution of the electric generating stations at as many points in the district as are required, and consequently it secures a great diminution of cost in the electric distributing mains. Von Rysselbergh also contemplated the production of a store of electricity at the sub-stations during hours when the demand for power or electricity was smallest.

The company formed to carry out von Rysselbergh's plans has erected a principal station in the Rue du Chantier, at Antwerp. Here are two compound steam-engines on the Sulzer

M

system, constructed by Messrs. Carels Frères. The cut-off is
variable for the high pressure cylinder, and regulated by the
governor. It is fixed for the low pressure cylinder. The high
pressure cylinders are jacketed with boiler steam, the low pressure
cylinders with steam from the intermediate reservoir. The
engines are arranged to work with or without condensation.

Diameter of high pressure cylinder . . .	.	27 ins.
„ „ low „ „ . . .	.	40 „
Stroke	.	$37\frac{1}{2}$ „
Revolutions per minute (normal) . . .	.	60 „

The speed can be regulated by the governor to anything
between 30 and 75 revolutions per minute. The speed is determined automatically, according to the demand for pressure
water. In addition, the accumulators govern the engines, so
that if at any time there is no demand for water the engines
are stopped, and start again automatically as soon as the flow
in the mains recommences. The discharge from the pumps is
13·7 gallons per revolution at $52\frac{1}{2}$ atmospheres. Two engines
are at present installed, but there is space for a third. The
pumps are driven by the prolongation backwards of the piston
rods. The pumps are on the Riedler system, with differential
action and valves closed mechanically. The pump plungers are
8 ins. in diameter. The accumulators are of special cast iron, with
rams 20 ins. in diameter, and 23 feet stroke. The engines, pumps
and accessories have been constructed by Messrs. Carels Frères,
the finish being perfect, and all the parts rigorously interchange-
able. The steam is supplied by five boilers, each having two
furnaces (internal) with corrugated flues and 16 'Galloway' tubes.
The steam pressure is 8 atmospheres. The boilers were constructed by the Société Anonyme de Chaudronnerie at Liège.

The distributing mains for the pressure water are steel pipes
of 11·8, 7·5, 5·85, and 5·15 inches diameter. These pipes are
all tested to 150 to 200 atmospheres. The flanges, T-pieces,
and junctions are of steel. The distributing mains will have an
extension of about $7\frac{1}{2}$ miles, and it is intended that there shall
be about 12 electric sub-stations. Each sub-station will distribute electricity over a radius of 1,600 feet by underground
conduits, which will have an aggregate length of 10 miles.

CHAPTER IX

TRANSMISSION OF POWER BY COMPRESSED AIR

COMPRESSED air has been employed in engineering operations for a long period, but it is only recently that its capabilities have been adequately recognised. The earliest important application of compressed air was for diving. Diving-bells are believed to have been used in the sixteenth century, and Smeaton in 1786 and Rennie in 1812 used them in important undertakings. Cubitt employed compressed air in sinking the piers of Rochester Bridge in 1851, and Brunel used a similar method at Saltash in 1854. Compressed air was used in driving the Thames Tunnel by Brunel, and Barlow employed it in the Thames subway. It has been largely used in tunnelling operations since that time. The shaft of the Marie colliery at Seraing was sunk by means of compressed air by the Société Cockerill in 1856.

The use of compressed air in transporting goods was suggested by Medhurst in 1810 and Vallance in 1818. Some early pneumatic railways were built. Similar methods have been adopted in recent years for transmitting messages and parcels in London and Berlin.

Papin seems to have considered the transmission of motive power by a vacuum method in 1688. Triger transmitted motive power by compressed air a distance of 750 feet at the mines of Chalonnes in 1845. Soon after compressed air was used in several collieries. The greatest impetus to the application of compressed air for transmitting power was due to its adoption for working the boring machinery at the Mont Cenis Tunnel. M. Sommeiller, in association with M. Kraft, made extensive experiments on compression at the Cockerill works at Seraing. On data so obtained the whole of the machinery for compressing, transmitting, and utilising compressed air at Mont

Cenis was designed and constructed at Seraing. At first a kind of hydraulic pneumatic ram was used for compression. In 1861, this was superseded by water-piston compressors driven by turbines. The air was transmitted a maximum distance of 20,000 feet to work the drills.

The air pressure used was seven atmospheres (105 lbs. per sq. in.). There were at Mont Cenis air motors worked expansively, the cylinders of which were heated externally to prevent freezing. In the construction of the St. Gothard Tunnel, in 1872, still more powerful air-compressing machinery was employed. The compressors were at first designed to be of small size, to run at a high speed, and to be cooled externally. But with a short stroke and quick speed there is not time for the heat developed by compression to be abstracted through the cylinder wall, and a spray injection suggested by Professor Colladon was added.

In 1877 Mekarski used air compressed to twenty-five or thirty atmospheres in conjunction with a small amount of high-pressure steam to drive tramway cars, and Mekarski was one of the first to use compound compressors.

In 1877 at Vienna and in 1881 at Paris, M. Popp installed a system for working and regulating a great number of clocks, by impulses of compressed air conveyed in pipes from a central station. A demand arose for a supply of the compressed air for working small motors, and this proved so successful that there has been developed in Paris the most important system of power distribution hitherto carried out. In Paris, motive power is transmitted to industries of every kind over a large area by air compressed at a central station, and even sub-stations for electric lighting are driven by air motors. It is interesting that in Paris a system of distributing motive power by vacuum, carried out by M. Boudenoot, has been successfully in operation since 1885. The motors are worked by atmospheric pressure, and exhaust into pipes in which a vacuum is maintained by air pumps at a central station. A system of pumping sewage at a number of scattered sub-stations by compressed air, supplied from a single compressing station, has been developed by Mr. Isaac Shone, and is in operation at several towns in this country and the United States, and at Rangoon.

Compressed-air transmission is a perfectly general method of distributing power for all purposes. Whether in any given case it is the most advantageous, the least wasteful of power, or the cheapest in working cost, depends on various circumstances. M. Hanarte believes that it is and will continue to be the most economical method of transmission to considerable distances.[1] The loss in the air mains is very small. The motors worked expansively are efficient. The mains can be carried by any path, and differences of elevation between the compressing and working points do not sensibly affect the result. In hydraulic transmission the water must be collected, stored, and in some cases filtered, and having actuated a motor, means must be found for removing it. But air is everywhere available, and can be discharged anywhere without causing trouble. Compressed air has peculiar advantages in the case of underground transmissions. It has been used to replace manual labour in situations where hardly any other motive power could have been employed. In driving a tunnel at a mine at Sacramento, for instance, the cost was reduced to one-half, and the rate of boring was three times as fast when compressed-air machinery replaced hand labour. In such cases the advantage is so great, even with uneconomical machinery, that the inducement to adopt very perfect machinery is absent. Hence much of the air-compressing plant at mines has been unnecessarily inefficient and wasteful of power. In many cases air-compressing plant has been driven by water power, and this also has tended to a neglect of the conditions necessary for economical working. Mr. Savage argues, with reference to the Terni Steel Works,[2] that the common objection to the use of compressed air on the ground of waste of power loses much of its force when the compressors are driven by an almost costless supply of natural energy such as water power. It is unfortunate for the reputation of the system of transmission by compressed air that the rough purposes to which it has been applied, the indifference to waste of power in mining and tunnelling operations, and the preference for simple and cheap machines, have delayed and hindered the improvement of compressed-air plant.

---

[1] 'Transmission du travail à distance par l'air comprimé'; *Congrès International de Mécanique Appliquée.* Paris, 1893.

[2] 'Terni Steel Works,' Savage ; *Proc. Inst. Civil Engineers*, vol. xciii.

A good deal was done to improve air compressors by Sommeiller, by Dubois and François, and by others, in the large plants constructed for Mont Cenis, for the St. Gothard works, and for some collieries. In the distribution of power in towns still further consideration has been given to the question of economy of working. But here again it has been very unfortunate that, in both the great installations in Paris and in Birmingham, there were conditions of development very unfavourable to the complete and fair trial of compressed air as a means of transmission. It is reasonably certain that with greater attention to scientific principles better results are attainable than have hitherto been reached in the use of compressed air.

For the special purposes to which power distribution is applied in London, the high pressure hydraulic system has great advantages. Where local conditions permit the construction of high level reservoirs, a system, like that in Zürich and Geneva, of hydraulic distribution is perfectly successful. But in more numerous cases compressed air is likely to prove preferable to hydraulic transmission. It is also the most important rival of electrical distribution. There are at present comparatively few cases where electrical distribution of power has been carried out, and though enough is known of the capabilities of electrical transmission to show that it could be adopted on a large scale with complete mechanical success, the cost of distribution of power by electrical methods is at present very imperfectly determined. For long distance transmission, and where cheap overhead conductors can be adopted. no doubt electrical methods have an important field of application. But up to the present time, and excluding transmissions for lighting, an enormously greater amount of power has been actually distributed by compressed air than by electricity.[1] So far as can be judged at present, in the case of distribution of power in towns, and especially where work has previously been done by steam-engines which can be converted into air motors, in such cases compressed air is likely to prove a more convenient and cheaper means of power distribution than electricity.

General Considerations on Compressed Air as a means of

---

[1] It has been stated by Professor Lupton that six makers of compressed air machinery in England have manufactured compressors working to an aggregate of 82,300 h.p.

Distributing Power in Towns.—The desiderata in a system of power distribution in towns may be shortly enumerated as follows :—

(1) The possibility of indefinitely subdividing the power distributed and measuring the supply to each consumer.

(2) Minimum first cost of distributing mains and minimum loss of energy in distribution.

(3) Simplicity, cheapness, and efficiency of the motors required by consumers of power, and especially that the motors should require little attendance and involve little risk.

(4) Freedom from danger to life or property when accidents occur to motors or distributing mains.

(5) Facility of adaptation to various requirements, additional to the supply of motive power. This is important, both from the additional revenue obtained, and because the more various the industrial applications satisfied, the better are the conditions of working at the central station. The fluctuations of demand are diminished, and the load-line improved.

A compressed air system meets these conditions on the whole more completely than any other system hitherto carried out. It has yet to be seen whether electrical distribution meets them equally well. Experience in Paris shows how great the facility is for subdividing the power in a compressed air system. There are motors ranging from 150 h.p. to less than one-tenth of a h.p. (45 foot lbs. per second). The majority of the air motors are, in fact, of less than one h.p. These can be started and stopped by merely opening or closing the supply valve, and the measurement of the air used presents no practical difficulty. In Paris and in Birmingham the air is measured by meters, which are not costly, and which are accurate enough to give satisfaction. As to the distributing mains, it may be pointed out that in an air system no return main is required, the air being discharged at the working point without creating any nuisance. In a steam distribution a return main is desirable, to avoid heat-loss, and in an electric distribution a return main is necessary. Air mains are less costly than hydraulic mains or steam mains. Under what conditions they are less costly than electric mains is a question yet to be determined. Probably they are less costly than electric mains, except in cases where high electric pressure can be used, and overhead conductors. M. Solignac has considered the case of the transmission of 70,000 h.p. from Billancourt to the Place de la Concorde at Paris, a distance of $4\frac{1}{4}$ miles.[1] He comes to the conclusion that air mains would cost 112,000*l.*, while electric mains worked at 2,000 volts would cost 700.000*l.*[2] Even if the energy were required at the terminus in the form of electricity, he concludes that it would cost 20 per cent. less to transmit by compressed air and generate electricity at the terminus, by dynamos driven by air motors, than to generate and transmit the electricity from Billancourt. As to loss of energy in the mains, electricity has little advantage over compressed air. The pressure loss in the mains

---

[1] Solignac, 'Transport de l'énergie par l'air comprimé ': *Congrès International de Mécanique Appliquée.* Paris, 1893.

[2] Estimates widely different from this will be found, especially in comparisons of the cost of electric and air transmissions made by electrical engineers. For instance, there is such a comparison in the report by Messrs. Zweifel and Hoffmann on a project for the electrical distribution of power in Mülhausen (*Mémoires couronnés par la Société Industrielle de Mulhouse,* November, 1892). The following is the comparison given :—

Let it be required to transmit 500 h.p. 2,000 metres. The cost of installation will be as follows :—

Air Transmission

	Fr.
For the compressor	12,000
Two thousand metres of main, 400 mm. in diameter, laying included	120,000
Air motors and re-heaters	65,000
	Fr. 197,000

Electric Transmission

	Fr.
For the dynamo.	50,000
Four thousand metres of cable, 120 mm. in section . .	32,000
Accessories and erection	8,500
Electric motors	40,500
	Fr. 140,000

It will be seen that for transmitting 500 h.p. an air main of 16 inches diameter is assumed ; but in Paris it has been shown that 90 cubic feet of air compressed to 5 atmospheres (450 cubic feet reckoned at atmospheric pressure) will yield one h.p. hour, in a not very perfect converted steam engine motor ; that is ·025 c. ft. of compressed air, or 0·125 c. ft. of air reckoned at atmospheric pressure, per second per h.p. But a main 16 inches in diameter, at velocities actually used, would transmit $1.4 \times 50 = 70.0$ c. ft. of compressed air per second : it would therefore carry 2,800 h.p., instead of the 500 which Messrs. Zweifel and Hoffmann assume. If their estimate of the cost of the air main is divided by five, the relative cost of compressed air and electric transmission will look very different.

of a town distribution is insignificant. In the Paris system the principal mains have an extension of 55,000 metres (34 miles). The loss of pressure between St.-Fargeau and the most distant point of the main rarely reaches 8 lbs. per sq. in. The safety of an air main is obvious, and even a leakage or burst of the main is much less serious and attended with less damage than that of a water or steam main. Air leakage is less dangerous than electric leakage. When an air distribution is introduced in a town, power users do not require new plant and need incur no outlay for motors. The boilers, with all their attendant disadvantages of stoking, removal of ashes, cleaning, and risk of explosion, are dispensed with, and the steam-engine, with little alteration, serves as an air motor. If an electric system is introduced, the old motors must be removed and new motors purchased. Further, if electric motors are themselves of high efficiency, they run at a high speed, and in most cases there is considerable loss in the gearing required to adapt them to ordinary purposes.

In regard to adaptability to various requirements, compressed air is in a very advantageous position. Electricity supplies power and light, but it cannot be used for supplying heat, except at a cost prohibitive in most applications. Gas supplies heat, power, and light. But for lighting it is open to obvious objections, and for heating and power it is expensive. Pressure water supplies power, and indirectly light, if a motor is used to drive a dynamo. But except where cheap water power is the original source of energy, it is too expensive for most purposes where motive power is required. Steam supplies heat, motive power, and indirectly light if a steam motor is used to drive a dynamo. But it is more expensive than compressed air, and involves more risk and attention. Compressed air can be supplied so cheaply that not only can it be used directly as a source of motive power, where that is the commodity required, but it can be advantageously used to drive sub-stations and private installations for generating electricity for lighting purposes or for working pumps and ventilating-fans. With a water-cushion between the air and the lift ram, compressed air is as convenient for working lifts as pressure water. It has been used in working cranes at the Cockerill works for forty years. Compressed air is not directly a source of heat, but used for blowing purposes it is an extremely useful adjunct to furnaces. In Birmingham, smiths' fires and cupolas have been worked direct from the air mains without any blowing machinery. A small jet of high pressure air induces a large stream at lower pressure. In Paris compressed air has important applications for refrigerating purposes. Besides large refrigerating stores in some restaurants, an air motor is used for driving a dynamo for lighting purposes, and the cooled exhaust from the air motor is used to cool chambers in which food is stored. Lastly, compressed air is already used in working tramways, and it appears likely that larger applications of this kind are possible.

General Arrangement of a System of Compressed Air Transmission.—The arrangements include—(1) A compressing plant driven by steam or water power, with air reservoirs of more or less capacity to diminish momentary fluctuations of pressure.

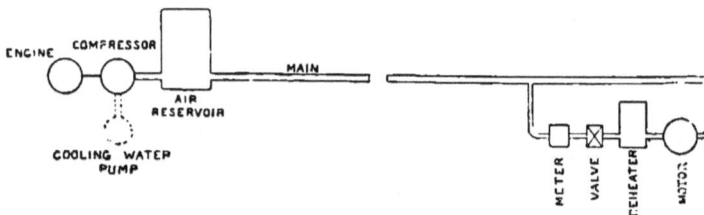

FIG. 46.—DIAGRAM OF PNEUMATIC SYSTEM.

The compressors usually require the addition of cooling arrangements for absorbing the heat developed in compression. (2) A system of air mains for distributing the compressed air to the working points. (3) Air motors driven by the compressed air, and sometimes provided with re-heaters, to increase the work done by the air and diminish the cooling during expansion. It is necessary, therefore, to consider the construction of compressors and their efficiency, the construction of mains and the losses in transmission, and the construction of air motors and their efficiency.

Fig. 46 shows diagrammatically the arrangement of a system of compressed-air transmission, with one motor system driven from the main. The circuit is, of course, an open one, no return main being used.

Properties of Air.—Let P be the pressure (absolute) in lbs. per sq. ft., V the volume of a pound in c. ft., and T the absolute temperature in Fahrenheit degrees. These quantities are connected by the relation

$$P V = 53 \cdot 2 \ T.$$

Let P_a V_a T_a be corresponding quantities for air at ordinary atmospheric temperature and pressure. It will be assumed that ordinary atmospheric air before compression is at a temperature of 60° Fahr. or 521° absolute, and at a pressure of 2,116·3 lbs. per sq. ft. For these conditions

$$P_a \ V_a = 53 \cdot 2 \times 521 = 27,710 \text{ ft. lbs.,}$$

and the volume of a pound of air is

$$V_a = 13 \cdot 09 \text{ c. ft.}$$

Action in the Compressor.—Consider for simplicity a single-acting compressor which receives and discharges a pound of air in each revolution, and let the effects of clearance and the resistances of the passages be neglected.

Let P_1, V_1, T_1 be the absolute pressure in lbs. per sq. ft., the volume of a pound in c. ft., and the absolute temperature of the air after compression;

P_a, V_a, T_a the same quantities for air before compression;

p_1 and p_a will be used for the corresponding pressures in lbs. per sq. inch;

$r = V_a / V_1$ is the ratio of compression, a quantity determined by the mechanical construction of the compressor;

$\rho = P_1 / P_a = p_1 / p_a$ may be termed the compression-pressure ratio. It depends on r, and also on the thermo-dynamic conditions of the compression.

Fig. 47 shows the indicator diagram of such a compressor. During the suction stroke a volume V_a at pressure P_a is drawn into the compressor cylinder. During compression to the volume V_1, the pressure rises according to some law expressed by the curve D C. Finally, the air is expelled into the mains at the pressure P_1. In general, the compression curve will lie between two curves D F, D G, corresponding to two limiting cases. If heat is abstracted from the air during compression, so that the temperature remains constant, the compression curve will be the isothermal D F defined by the relation

$$P V = \text{constant}$$
$$= P_a \ V_a = 27,710.$$

If no heat is added or abstracted during compression the temperature of the air will rise, and the compression curve will be the adiabatic D G defined by the relation

$$P V^\gamma = \text{constant},$$

where $\gamma = 1 \cdot 41$.

In ordinary compressors the curve lies between D F and D G, and approximates sufficiently to a curve defined by the relation

$$P V^n = \text{constant},$$

n having a value between 1 and $1 \cdot 41$.

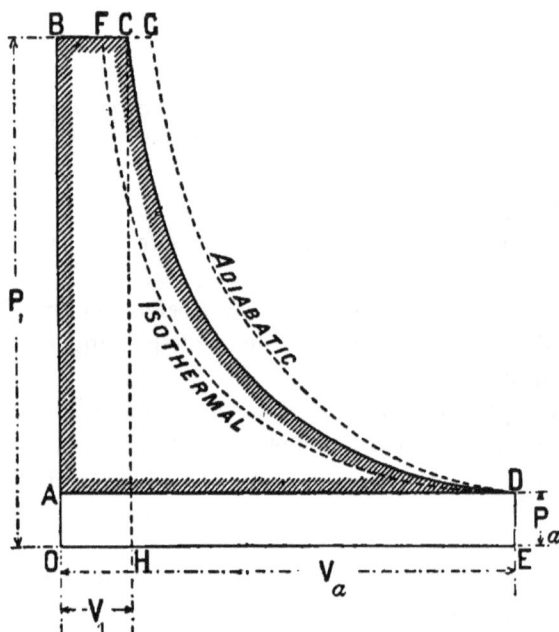

Fig. 47.

The whole work of a complete double stroke consists of three parts :—(1) The work O A D E of the atmosphere on the piston during the suction stroke; (2) the absolute work of compression E D C H; (3) the work of expulsion of the air into the mains H C B O. The effective work is the sum of these

$$- O A D E + E D C H + H C B O,$$

that is, the shaded area A B C D.

Case of Isothermal Compression.—It will be shown presently that the most economical compressor mechanically would be one in which heat is abstracted during compression, so that the compression is isothermal. In that case the effective work is (fig. 48). since $PV =$ constant,

$$- P_a V_a + P_a V_a \log_e \frac{P_1}{P_a} + P_1 V_1$$

$$= P_a V_a \log_e \frac{P_1}{P_a},$$

or exactly equal to the absolute work of compression H F D E. But the heat abstracted during compression is equal to the same

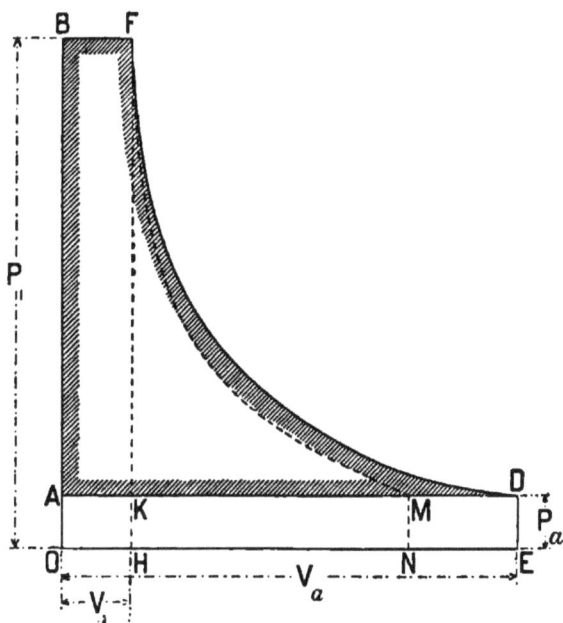

Fig. 48.

quantity. Hence the curious result is arrived at that in the most economical compression, the effective work of compression is entirely abstracted as heat and wasted. All the compression does is to put the air in a condition to do work in a motor at the expense of its intrinsic energy. In that way there is obtained an amount of work nearly equal to the work done in compression. But the work in the motor is not strictly the restoration of the energy expended in the compressor, but

174 DISTRIBUTION OF POWER

energy borrowed from the air. Hence the conditions of transmission of power by compressed air are different from those of transmission by pressure water.

Case of Adiabatic Compression.—The volume of one pound of air at the final pressure P_1 will be

$$v_1 = v_a \left(\frac{P_a}{P_1}\right)^{\frac{1}{\gamma}} = 13.09 \left(\frac{P_a}{P_1}\right)^{0.71}$$

The absolute work of adiabatic compression is per pound of air

$$\frac{P_1 V_1 - P_a V_a}{\gamma - 1}.$$

Hence the effective work in one revolution of the compressor (A B G D, fig. 47) is

$$\frac{P_1 V_1 - P_a V_a}{\gamma - 1} + P_1 V_1 - P_a V_a$$

$$= \frac{\gamma}{\gamma - 1} P_a V_a \left[\left(\frac{P_1}{P_a}\right)^{\frac{\gamma-1}{\gamma}} - 1 \right]$$

$$= 95630 \left[\rho^{0.29} - 1\right] \text{ foot lbs.}$$

Case of Partially Cooled Compression.—The general equation for the work expended in compression is the same as for adiabatic compression, if n is substituted for γ.[1] If the index of the expansion curve is $n = 1.25$, the work expended per pound of air is

$$138550 \left[\rho^{0.2} - 1\right] \text{ foot lbs.}$$

Rise of Temperature during Compression.—For isothermal compression the temperature is constant. In any other case

$$T_1 = 521 \rho^{\frac{n-1}{n}}$$

For adiabatic compression substitute γ for n. The rise of temperature is considerable, as the following table shows:—

[1] Let $P_a V_a T_a$ correspond to the initial, and $P_1 V_1 T_1$ to the final, conditions in any compressor. Then

$$n = \frac{\log (P_1/P_a)}{\log (V_a/V_1)}$$

$$= \frac{\log (P_1/P_a)}{\log (P_1/P_a) + \log (T_a/T_1)}$$

$\frac{P_1}{P_a}$	p_1 lbs. per sq. in. absolute	Temperature reached in compression		
		Isothermal $n=1$	Partially cooled $n=1.25$	Adiabatic $n=1.41$
2	29.4	60°	137°	176°
3	44.1	60	187	255
4	58.8	60	226	318
5	73.5	60	256	370
6	88.2	60	284	415
7	102.9	60	307	455

Unnecessary Waste of Work in Heating the Air in the Compressor.—If the compressed air were used in a motor directly adjacent to the compressor, in its heated state, there would be no necessary loss due to rise of temperature during compression. Commonly the air is used at a distance, and has cooled from T_1 to atmospheric temperature T_a and shrunk in volume from B C to B F (fig. 47) before reaching the working point. The most economical compression for air transmission would be isothermal compression. The area F D C represents work expended in the compressor which is wasted before the air is used.

It can be shown that the work wasted in heating the air in the compressor above its initial temperature, when the expansion curve is given by the relation

$$P V^n = \text{constant},$$

is,—
$$P_a V_a \left\{ \frac{n}{n-1} \left[\rho^{\frac{n-1}{n}} - 1 \right] - \log_e \rho \right\}$$

$P_a V_a$ for air taken from the atmosphere $= 27710$. The work wasted is given in foot pounds per lb. of air compressed.

WORK WASTED IN HEATING IN COMPRESSOR PER POUND OF AIR
COMPRESSED

$\frac{P_1}{P_a}$	p_1 = Pressure of compressed air in lbs. per sq. in.		Work wasted	
	Absolute	By gauge	Adiabatic compression $n=1.41$	Partially cooled compression $n=1.25$
2	29.4	14.7	0.077 $P_a V_a$	0.052 $P_a V_a$
4	58.8	44.1	0.322 "	0.209 "
6	88.2	73.5	0.562 "	0.363 "

It will be seen that the loss increases rapidly, almost as the square of the ratio P_1/P_a. This rapid increase of heating loss has led many constructors to advise the use of very low working pressures in compressed-air transmissions. But that involves an oversight. The increased loss at the compressor, due to a higher working pressure, is partly balanced by an increased efficiency of the air motor, so that low working pressures are not necessarily most economical for the whole system.

Efficiency of Compressor and Motor Combined.—If the air is compressed isothermally, as it very nearly is in modern compound compressors, then the question arises how much work is obtained in the compressed air motor. If the air is used non-expansively, as in the smaller motors used in Paris, and in much compressed-air plant used for rock-drilling and similar purposes, the work done in the motor, neglecting valve resistances and friction, is simply the admission work $P_1 V_1$, less the work of expulsion, $P' V_1$. That is, the work obtained is

$$P_a V_a \left(1 - \frac{P_a}{P_1} \right)$$

Then the efficiency of combined motor and compressor, with the most perfect system of compression, but with non-expansive motors, and neglecting all subsidiary losses due to clearance, resistance in mains, and mechanical friction of compressors and motors, would be

$$\eta = \frac{1 - \frac{P_a}{P_1}}{\log_e \frac{P_1}{P_a}} = \frac{\rho - 1}{\rho \log_e \rho}$$

ρ or $P_1 =$ 2   3   4   6   8   10 atmospheres
$\eta =$ 0·72   0·61   0·54   0·46   0·42   0·39 ;

the practically realised efficiency would be, of course, materially less than this.

The efficiency diminishes rapidly with increase of initial pressure, and it is the use of bad and inefficient air plants of this kind which has given compressed air a bad name, and made engineers hesitate to adopt high working pressures. To obtain good results the air must be used expansively.

Suppose the air compressed isothermally and then expanded

in the motor down to atmospheric pressure without gain or loss of heat, nearly realisable conditions, and let clearance and loss in mains be neglected, as well as friction. Then the indicator diagram of the motor (fig. 48) will be A B F M, the expansion curve being an adiabatic F M, and the air cooling during expansion. The efficiency of the whole arrangement will be the ratio of the area A B F M to the area A B F D. It is easy to show that that efficiency will be given by the equation

$$\eta = \frac{95,600 \left[1 - \left(\frac{P_n}{P_l}\right)^{0 \cdot 29} \right]}{27,710 \log_e \frac{P_l}{P_n}}$$

RESULTANT EFFICIENCY OF COMPRESSOR AND MOTOR

Working pressure, lbs. per sq. in.		$\frac{\eta}{p_a}$	η
By gauge	Absolute p_l		
14·7	29·4	2	·906
44·1	58·8	4	·824
73·5	88·2	6	·780
102·9	117·6	8	·752

It is obvious from the diagram that, even if the ratio of expansion in the motor is not quite great enough to reduce the air to atmospheric pressure, the loss of work is not very great.

Practically, it is necessary to use gauge pressures of 45 lbs. per sq. in. at least, that the machinery may not be too cumbrous. The calculation shows that when using the air properly much higher working pressures may be adopted without serious increased loss of efficiency.

Methods of Cooling the Air during Compression.—Various means have been adopted to cool the air during compression, and so to reduce the unnecessary heating loss. Sommeiller adopted water pistons, the air being displaced by a mass of water driven by an ordinary solid piston. The water directly absorbs part of the heat, and the cylinder walls are kept cool and moist. In the early compressors of this type very good results as regards efficiency were obtained. But the Mont Cenis compressors were worked at a slow piston speed, and were

N

cumbrous and expensive for the amount of work done. It may
be stated, however, that Mr. Leavitt still uses water piston
compressors at the Calumet mines, and by giving suitable form
to the pistons and passages he succeeds in running them at a
good speed, and obtains a very satisfactory efficiency. M.
Hanarte also, in France, has constructed water piston compressors,
with paraboloidal chambers in which the energy of the water
pistons is quietly diminished, and shock and loss of energy are
avoided.

At the St. Gothard Tunnel works compressors of a less
effective type were adopted. The compressor cylinder was dry,
but was surrounded by a water-jacket. Air does not part with
heat readily to a metal surface, and the heat produced was very
imperfectly abstracted. Later a method proposed by Professor
Colladon was adopted. ˙Water was injected in a fine spray into
the cylinders, and a much more powerful cooling action was obtained. To some extent, however, the cooling is deceptive, as
it takes place late in the stroke and during discharge of the air,
and the compression curve is not lowered as much as it
should be.

In order that the expansion may be isothermal, an amount
of heat must be removed equal to the work of compression.
That is—

$$P_a V_a \log_e \frac{P_1}{P_a} \text{ foot lbs.}$$

$$= 35 \cdot 61 \log_e \rho \text{ Th. U.}$$

If there is only a small rise of temperature, the heat removed
will not be very different. Suppose the injection water received
at 60° and discharged at 100°. Each pound will remove 40
Th. U. Hence the amount of injection water in pounds per
pound of air compressed must be

$$0 \cdot 885 \log_e \rho.$$

$\rho = p_1/p_a$	Pounds of injection water per pound of air
2	·61
4	1·23
6	1·59
8	1·84
10	2·04

Compound Compressors.—It has been shown that the heating of the air during compression is a serious cause of loss of efficiency, and that the means adopted to cool the air during compression by water jackets or spray injection are not perfectly satisfactory. A more efficient means of approximating to isothermal compression is to carry out the compression in stages, in a compound compressor, and to cool the air to atmospheric temperature in intercoolers, between each stage of the compression. Compound compressors were first used by Mekarski and others in cases where high pressures were required. Mr.

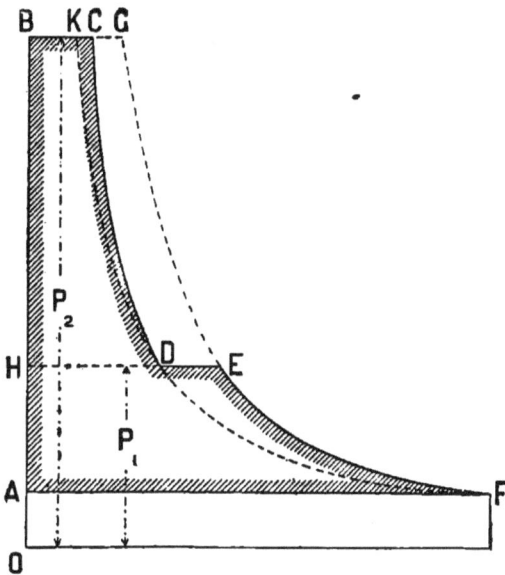

FIG. 49.

Northcott made a compound compressor with intercooler in 1878. The Norwalk Company in America constructed compound compressors for mines in 1880, apparently at first with a view of equalising the effort during the stroke. In 1881, however, they introduced an intercooler between the high and low pressure cylinders. This was made like a surface condenser, with thin brass pipes through which cooling water circulated.

The effect of the intercooler is very important in reducing the heating loss. The compound diagram is shown in fig. 49. Air is compressed from p_a to p_1 in the low pressure cylinder. Then

in passing through the intercooler it shrinks in volume from
H E to П D, the point D being on the isothermal, if the cooling
surface is sufficient.   It is then further compressed to p_2 in the
high pressure cylinder.   The work saved by intermediate cooling is
the area E D C G.   The adoption of these intercoolers is important,
because it removes the chief objection to the adoption of high
working pressures in air transmissions.

Let T_a be the absolute temperature at which air is admitted
to either cylinder; T_1 and T_2 the temperatures at the end of each
stage of compression; p_a, p_1, p_2, the corresponding pressures in
lbs. per sq. in.   Then, if the compression is adiabatic,

$$\frac{T_1}{T} = \left(\frac{p_1}{p_a}\right)^{0.29}$$

$$\frac{T_2}{T_a} = \left(\frac{p_2}{p_1}\right)^{0.29}$$

$$T_1 T_2 = T_a^2 \left(\frac{p_2}{p_a}\right)^{0.29} = \text{constant.}$$

The final temperature will be lowest if $T_1 = T_2$, whence—

$$p_1 = \sqrt{p_2 p_a},$$

which determines the proportions of the cylinders.   The total
work of compression for the two stages is per pound of air—

$$95,630 \left\{ \left[\left(\frac{p_1}{p_a}\right)^{0.29} - 1 \right] + \left[\left(\frac{p_2}{p_1}\right)^{0.29} - 1 \right] \right\}$$

$$= 191,260 \left[\left(\frac{p_2}{p_a}\right)^{0.145} - 1 \right]$$

The ratio of the volumes of the cylinders, if the above condition
is satisfied, is simply

$$p_1 / p_a = \sqrt{(p_2/p_a)}.$$

The Influence of Clearance Space in Compressors.—The clearance space diminishes the amount of air delivered, so that a
larger compressor is necessary for a given output.   The air compressed in the clearance space to p_1 expands as the piston
returns, and prevents the opening of the suction valves till its
pressure has fallen to p_a.   There is no direct diminution of
efficiency, because the work of compressing the clearance air is
given back in the return stroke; but indirectly it diminishes

efficiency, because the larger the compressor the more work is wasted in friction.

Let v be the volume described by the piston per single stroke, cv the clearance volume. The expanding air takes heat from the cylinder walls, so that the expansion is nearly isothermal. Then the clearance air cv at p_1 expands to $(p_1/p_a)\, cv$ $= \rho cv$ during the first part of the return stroke. The amount of air in the cylinder at the end of the suction stroke is $v\,(1+c)$, so that the amount of air which enters and is delivered per stroke, reckoned at atmospheric pressure, is—

$$v\,(1+c)\ -\rho cv$$
$$=v-cv\,(\rho-1).$$

The ratio of volume delivered to volume described by piston or volume efficiency is—

$$\mu=1-c\,(\rho-1).$$

This diminishes as p_1/p_a increases, and becomes zero at some pressure given by the relation

$$\rho=\frac{p_1}{p_a} = 1 + \frac{1}{c}.$$

For instance, if the clearance is 10 per cent., the compressor will furnish no air if worked at 11 atmospheres, and only half the volume described by the piston if worked at 6 atmospheres. The importance of reducing clearance is therefore obvious. With solid pistons in dry compressors some clearance is necessary ; but it may be reduced to 1 to 3 per cent. of the volume described by the piston, by careful design. With water-piston compressors the clearance may be reduced to zero, though at high pressures an equivalent source of loss would arise from air absorbed by the water during the compression stroke, and given off during the return stroke. In compound compressors the effect of clearance is diminished, because for each cylinder p_1/p_a is less than in a simple compressor.

Loss due to Delayed Valve Action.—If the valves close too late, there is reflux of air into the cylinder, or from the cylinder into the atmosphere, producing a diminished volume efficiency. The loss is similar to that due to clearance. If the valves open late, there is an excess pressure at the latter part of the compression stroke, or a diminished pressure during the suction stroke, either involving a loss of work. With automatic valves some loss of this kind is unavoidable. With mechanically moved valves there is also often some loss from the ports opening and closing too gradually. Professor Riedler's valves have an advantage in this respect. Generally, in compressors the suction passages have an area of one-quarter to one-sixth that of the piston, and the delivery passages of one-tenth to one-twelfth.

Loss due to Heating Air during Admission.—In some compressors the air takes heat from the walls of the admission passages. The effect is to diminish the weight of air admitted per stroke, so that the volume efficiency of the compressor is diminished.

It is desirable that the air taken into the cylinder should be taken from a place where the air is as cold as possible. The following table gives the diminution of output for different values of the temperature of the air in the cylinder at the end of admission :—

Temperature of air at end of suction stroke	Volume of air delivered reckoned at 60° and atm. pressure	Loss of volume delivered due to heating of air admitted per cent.
60°	100	—
75	97·5	2·5
90	94·9	5·1
110	91·6	8·4

The following table gives the results of some experiments on the delivery of a 'Dubois-François' compressor, with cylinder 18 inches diameter, 48 inches stroke :—

Piston speed ft. per min.	Revs. per min.	Air delivered at atm. pressure and temperature in per cent. of volume described by piston
80	10	94
160	20	92
200	25	90
240	30	86
280	35	78

Compressor Valves.—In most compressors simple automatic or fluid-moved valves are employed. The objection to them is, that they create a resistance to the passage of the air, which becomes very serious at high speeds. If they are loaded so as to close quickly by springs, they do not open enough to let the air pass without loss of pressure. In many compressors mechanically-moved valves are employed; for instance, slide or 'Corliss' valves moved by an eccentric, or better by cams. The compressor can then be run faster, but there is still usually

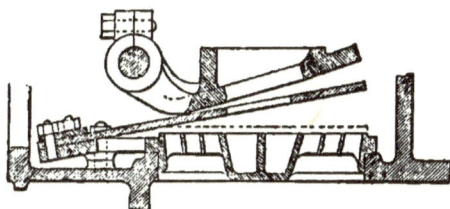

Fig. 50.

resistance, due to the valves opening and closing too slowly. Professor Riedler uses valves which open automatically and are closed mechanically. Fig. 50 shows one form of these valves. This is a flap valve, so set that it has a slight tendency to open. It is closed by a lever worked by a cam.

The following table, giving dimensions of the 'Norwalk' compressors, may serve as a guide to ordinary proportions:—

TABLE OF DIMENSIONS OF 'NORWALK' COMPRESSORS

Diam. of low pressure air cylinder ins.	Diam. of high pressure air cylinder ins.	Diam. of steam cylinder ins.	Length of stroke ins.	Revolutions per min.	Horse-power	Capacity in cubic ft. per min.	Steam pipe ins.	Exhaust pipe ins.	Air pipe ins.	Water pipe ins.
8	5	8	10	200	15	116	2	$2\frac{1}{2}$	2	$\frac{3}{4}$
10	$6\frac{3}{4}$	10	12	190	28	207	$2\frac{1}{2}$	3	$2\frac{1}{2}$	$\frac{3}{4}$
14	$9\frac{1}{2}$	14	16	150	55	427	3	4	4	1
20	$13\frac{1}{3}$	20	24	110	125	960	5	6	5	$1\frac{1}{4}$
26	$17\frac{1}{4}$	24	30	90	215	1,659	6	8	6	$1\frac{1}{4}$
32	$21\frac{1}{2}$	30	36	80	350	2,686	7	10	8	$1\frac{1}{2}$

These are adapted for air pressures of 60 to 100 lbs. per sq. in.

Mr. Pearsall's Hydraulic Air Compressing Engine.—Mr. H. D. Pearsall, apparently incited by a study of the early hydraulic ram compressors of Sommeiller, has designed a very interesting hydraulic compressing engine, in which air is compressed directly by a water-column without cylinders and pistons.

Fig. 51 shows this engine in section. A is the supply pipe by which water, with the energy due to its descent from a higher level, flows into the apparatus. C is a large cylinder valve which, when open, allows the water to flow out into the tail-race, and, when shut, forces the column of water to rise into the compression chamber M. The column of water in the supply pipe is allowed to acquire velocity by outflow into the tail-race. The valve C is then mechanically closed, and the descending column expends its energy in compressing the air in the chamber M, and discharging it into the receiver E. The cylinder valve C is actuated by a small air motor. The chamber M empties the water through the cylinder valve C, and fills with air through the air valve H, which is controlled by a float, K. The adjustment is such that the column of water can be made to come to rest at the instant when it reaches the delivery valve-plate.

Mr. Pearsall claims that very high velocities of flow can be permitted without danger or loss of efficiency. Some experiments made with a small apparatus gave an efficiency of 80 per cent.

The engine is simple, and there seems no reason why it should not have a high efficiency. But, till experiments have been made on a larger scale, it is impossible to say what the delivering capacity of the machine in a given time is. Till that is determined, it is uncertain whether it would be more costly, or less costly, than ordinary compressors worked by turbines.

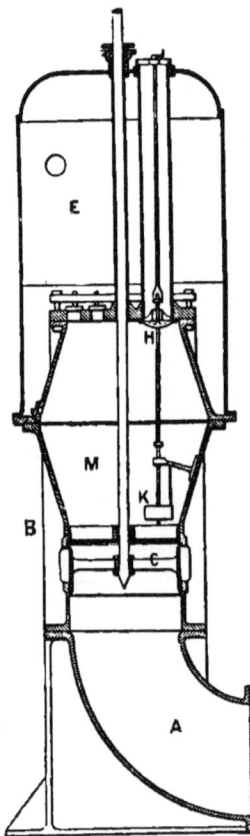

FIG. 51.

Losses in Transmission.—The frictional resistances in a pipe conveying fluid are proportional to the density of the fluid. Consequently at equal velocities the frictional resistance of air is enormously less than that of water. Conversely, air may be transmitted in air mains without serious fall of pressure at ten or twenty times the velocity practicable with water in water mains. Air at 90 lbs. per sq. in. pressure is about 115 times lighter than water, and the frictional resistance at equal velocities is less than one per cent. of that of water. In air mains there is nothing analogous to the hydraulic shock due to changes of velocity, which, as well as the friction, leads to a limitation of the velocity of water in mains to 3 feet per second in most cases, or to 6 feet per second in some cases.

In air mains velocities of 30 to 50 feet per second are allowed without serious frictional loss. In consequence of this high velocity, large amounts of power can be transmitted by air at moderate pressures, and in mains of moderate dimensions.

Most of the hydraulic mains of the London Hydraulic Power Company are 6 inches in diameter, the pressure is 750 lbs. per sq. in., and the velocity 3 feet per second. That corresponds to the transmission of 90 effective h.p. by each main. But air at 45 lbs. pressure per sq. in., with a velocity of 50 feet per second, would transmit 150 effective h.p. in a main of the same size. The largest high pressure hydraulic mains are $7\frac{1}{2}$ inches in diameter, and will transmit 140 h.p. But there is hardly any limit to the size of air mains, or the amount of power they will carry. The new Paris main from the Quai de la Gare to the Place de la Concorde, 20 ins. diameter and 7 kilometres in length, with air at 90 lbs. pressure per sq. in., transmits 6,000 h.p.

In the older Paris mains, which were carried through the sewers, and which had an exceptionally large number of bends, draining boxes and other sources of resistance, the frictional resistance, with a velocity of 25 to 30 feet per second, amounted to 2 lbs. per sq. in. per mile of main. It would be only in very long distance transmissions that the fall of pressure in the mains would be large enough to sensibly affect the efficiency of the system.[1]

As to the precise way in which a fall of pressure in the mains influences the efficiency there is a word to be said. If air enters an air main at 60 lbs. per sq. in. gauge pressure and

[1] M. Solignac says: 'Dans le réseau de la Compagnie Parisienne, qui compte 55,000 mètres de développement pour la force motrice, la perte de charge entre Saint-Fargeau et le point extrême de la conduite est à peine d'un kilogramme' (7 lbs. per sq. in.). This refers to ordinary working and to a main forming a network. The statement above gives the result of a direct experiment on a continuous main.

reaches the other end at 55 lbs. gauge pressure, there being a
fall of pressure of 5 lbs. due to friction, then it is commonly
stated that five-sixtieths of the energy of the air is wasted.   But
this is altogether erroneous, the statement being based on a false
hydraulic analogy.   With the fall of pressure in the case of air
there is an expansion of volume which largely compensates for
the loss of pressure.   The intrinsic energy of the air from which
the work of the air motor is borrowed remains constant.   It is
only because the air motor works against the pressure of the
atmosphere, that the available energy of air at 55 lbs. is less
than that of air at 60 lbs. pressure.

Suppose that a given amount of work can be done by an air
motor using a ton of air at 60 lbs. gauge pressure, or 75 lbs.
absolute.   The work expended in driving this motor, by a compressor adjacent to it, would be the work of compressing one
ton of air to 75 lbs. absolute pressure.   Now let the compressor
be removed to a couple of miles' distance, and the air supplied to
the motor through a main in which there is a fall of pressure of
5 lbs. per sq. in.   To do the same work as before, all that is
necessary is that a ton of air should be compressed to 80 lbs.
absolute pressure.   The difference between the work of compressing a ton of air to 80 and to 75 lbs. absolute pressure is the
loss of work arising out of the friction of the main, though it is
not rightly described as energy wasted in friction in the main.
This amounts with fairly good compressors to about 3 per cent.

In comparatively short distance transmissions, such as those
in towns, the loss of pressure in the mains is so insignificant
that it may be neglected.   In long distance transmissions an
accurate estimate of the frictional loss is necessary.   The author
believes that he has shown, using data derived from careful
experiments on twenty miles of main in Paris, that long distance
transmission of power by compressed air is perfectly practicable.
It is possible, with compressors driven by engines working to
10,000 i.h.p., to transmit the air in a main of not unusual size
a distance of twenty miles, and to obtain in motors worked by
the transmitted compressed air from 4,000 to 5,000 i.h.p., if the
air is used cold ; or 6,000 to 7,000 i.h.p., if the air is re-heated
before use.

Air Mains.—The air mains are ordinary wrought-iron tubing
for small sizes, and steel or cast-iron pipes for large sizes.   M.

Solignac gives the following as the dimensions and cost of the air mains in Paris.

Diameter. Inches	Cost per yard. Shillings	
1·6	4·74	
2·4	5·84	Wrought iron
3·2	6·72	
4·0	9·12	
12·0	24·10 (in sewer)	Cast iron
12·0	21·90 (in earth)	

It will be seen that cast-iron mains have been chiefly used in Paris. The new 20-inch main is a steel main. In Birmingham the large mains were of riveted wrought-iron. The principal difficulty is the jointing of the lengths of main. For cast-iron pipes, and for the Birmingham wrought-iron mains, sockets with lead joints have been adopted. But the variations

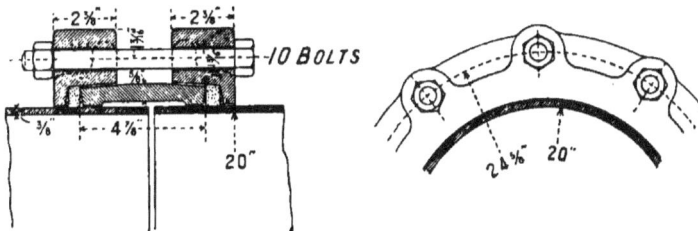

FIG. 52.

of temperature in an air main are liable to be greater than in water mains. In Birmingham the variations of temperature were excessive. The expansion and contraction makes lead joints leaky, and the leakage at Birmingham is said to have amounted to 45 per cent. of the whole air supply. At Portsmouth flanged joints are used, and they are said to give no trouble, but the distance of transmission is comparatively small. The only joint quite suitable for air supply is one adopted in Paris, and which is of a type described in Reuleaux's Constructeur as the ' Normandy ' joint. The lengths of main have perfectly plain spigot ends. A kind of double stuffing-box is formed over the two pipe ends, the packing consisting of two india-rubber rings. Fig. 52 shows the Paris joint. The india-rubber is compressed by two rings drawn together by bolts, and is so protected from access of air or water that it appears to be nearly imperishable. Perfect freedom of expansion and contraction is secured at each joint in the main. In the case of the Paris main (except some parts constructed in the sewers at an early date) the leakage is almost negligible. This joint appears to have been perfectly satisfactory in Paris, and has been adopted at Offenbach and in other German systems of compressed air distribution.

AIR MOTORS

The air motor is essentially a reversed compressor. If it works without expansion, it is inefficient; if it works with expansion, the air cools and there is, in some cases, trouble from the formation of ice in the valve passages. Rounding the edges of the ports so that the ice can be pushed away diminishes the trouble. Heating the air before it enters the motor, or during expansion, entirely obviates it.

Cooling during Expansion in the Motor.—In order to show how important the cooling action during expansion in the motor cylinder is, temperature curves have been drawn in fig. 53 for expansion from various pressures down to atmospheric pressure, that is for what may be termed complete expansion. The full curves show the fall of temperature during adiabatic expansion, for which

$$p v^{1\cdot41} = \text{constant}$$

is the law of expansion. The dotted curves show the corresponding fall of temperature when some heat is supplied during expansion, so that

$$p v^{1\cdot25} = \text{constant}$$

is the law of expansion. If $p_1 v_1 T_1$ are the initial pressure, volume of a pound, and absolute temperature, and $p_2 T_2 v_2$ the corresponding quantities at any stage of the expansion, then

$$\frac{T_2}{T_1} = \frac{p_2 v_2}{p_1 v_1} = \left(\frac{p_2}{p_1}\right)^{\frac{n-1}{n}}$$

where n has the values $1\cdot41$ or $1\cdot25$, according as the expansion is adiabatic, or with heat supplied as assumed above.

The lower set of curves show the temperature fall when the air is initially at a temperature of 60° Fahr. ; the upper set of curves, the temperature fall when the air is reheated before use to 200° Fahr. The following table gives the principal results. The temperature fall is given when the expansion is complete, so that the terminal pressure in the motor is one atmosphere. Suppose that on account of ice difficulty, or for any other reason, it is desired that the terminal temperature should not be below 32° Fahr., then the table gives the limiting terminal pressure possible under that condition.

The curves and the table show strikingly how limited is the possible range of expansion, with air not re-heated before use, if

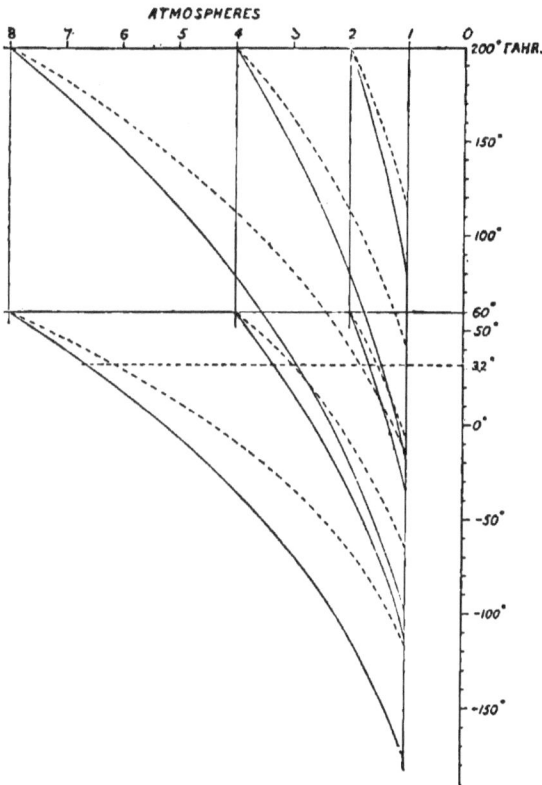

FIG. 53.

the difficulty of a very low temperature of exhaust is to be avoided. On the other hand, they show how much the possible range of expansion is increased by moderate re-heating of the air before use, when the exhaust does not fall below freezing point. The effect of some heat supply during expansion, sufficient merely to alter the index of the expansion curve from 1·41 to 1·25, is not very great on the terminal temperature.

| ADIABATIC EXPANSION | | | PARTIALLY-HEATED EXPANSION | | |
| *Initial Temp.* 60° | | | *Initial Temp.* 60° | | |
Initial pressure, atmospheres	Temp. after expansion to one atm.	Pressure after expansion to 32° F.	Initial pressure, atmospheres	Temp. after expansion to one atm.	Pressure after expansion to 32° F.
2	− 35° F.	1·65	2	− 8° F.	1·5
4	−113	3·3	4	− 66	3
8	−176	6·6	8	−117	6·02
Initial Temp. 200°			*Initial Temp.* 200°		
2	80° F.	Complete	2	114° F.	Complete
4	− 19	1·45	4	40	Complete
8	− 99	2·90	8	−15	1·82

Types of Motors.—Some special rotary motors are used in Paris for very small powers. They are not economical, and are rather costly. Most air motors are simply non-condensing steam engines, working with air instead of steam. In such converted machines there is often not inconsiderable loss from leakage, especially piston leakage. In a steam engine the condensation on the cylinder wall helps to make the piston tight. Air motors are not in the same position, and extra care should be taken to prevent leakage.

Several methods may be adopted to diminish the cooling difficulty and to increase the efficiency of the motor; one is to inject warm water into the motor cylinder in spray. The air takes heat from the water during expansion, reducing it in temperature to 32°. The good effect due to injecting even moderately warm water is considerable. The following short table gives the heat which must be given to the air to keep its temperature, during expansion down to atmospheric pressure, from falling below 32°; also the number of pounds of water which must be supplied per pound of air, if only the sensible heat of the water is utilised :—

Absolute pressure of air, lbs. per sq. in.	Heat required per pound of air in Th. U.	Pounds of water per pound of air, water supplied at		
		75°	100°	150°
29·4	53	1·23	·78	·45
58·8	106	2·46	1·56	·90
88·2	137	3·20	2·00	1·16
117·6	160	3·72	2·35	1·36
147·0	176	4·10	2·60	1·50

Injection of Steam into the Cylinder of the Air Motor.—Steam has an enormous advantage over warm water as a means of diminishing cooling during expansion, because it enters in a form extremely convenient for distributing the heat throughout the mass of air, and because the steam gives up its latent heat to the air. If we suppose the steam merely at atmospheric pressure, which is accurate enough for an approximate calculation, then each pound of steam received at 212° and rejected at 32° will give up 1,146 Th. U. Hence, making the same calculation as before, the following table gives the weight of steam which must be injected per pound of air to prevent the temperature in the motor cylinder from falling below freezing point.

Absolute pressure of air in lbs. per sq. in.	Heat required per pound of air in Th. U.	Pounds of steam required per pound of air
29·4	53	·046
58·8	106	·093
88·2	137	·120
117·6	160	·140
147·0	176	·154

Re-heating the Air before Use.—If the air is heated before entering the motor, the practical difficulty due to the cooling in expansion is entirely obviated, and an increase of the efficiency of the motor is obtained, which is of the greatest economical importance. The air when heated expands, and less air is used per stroke. Whether it is economical or the reverse to heat the air before use depends on this: whether the additional work obtained is more or less valuable than the coal expended. Experience shows that it is extremely advantageous economically to re-heat the air. The heat supplied is used with great efficiency, and a larger fraction of it is converted into work than in ordinary heat engines. Further, very small, easily managed, and simple re-heating apparatus can be employed. A simple coil of pipes, with a small furnace capable of heating the air current 300° Fahr., may increase the work done per pound of air by 25 or 30 per cent. The heat, according to the experiments of Riedler and Gutermuth, is used five or six times as efficiently as heat supplied to a good steam engine.

Fig. 54 shows a simple form of re-heating oven. The compressed air passes through a double spiral pipe c. The furnace

gases rise through the centre of the coil and descend on the outside in a cast-iron casing, with a spiral diaphragm or rib. The
grate is at F. The air discharged from the motor M may be used
to create a chimney draught. This has the advantage that the
draught varies with the amount of work done. As air does not

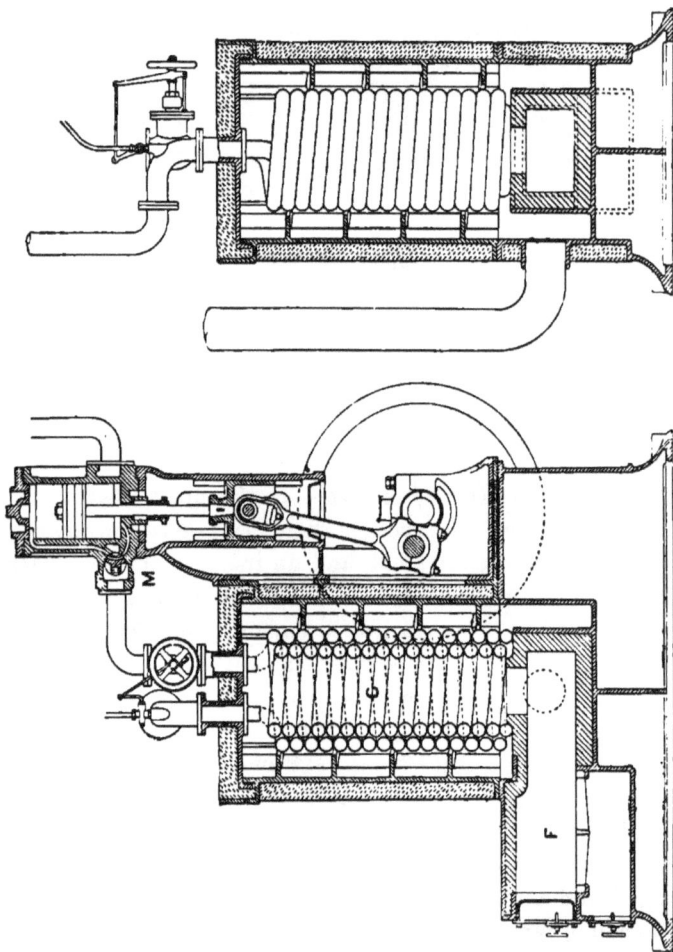

FIG. 54.

readily take up heat from metal surfaces, it is advantageous to
introduce a small quantity of water into the spiral pipe of the
re-heater. The water is evaporated into steam, and in the motor
the steam condenses, giving back its latent heat to the expanding
air. The water may be supplied from a reservoir above the

oven, to which the air pressure is admitted so that the water descends into the heater by gravity. The reservoir can be re-filled by shutting off the air pressure. Steam thus used is extremely efficient in increasing the work done by the air, and probably the moisture in the cylinder helps to prevent wear and leakage.

In some simple re-heaters tested by Professor Gutermuth in Paris, the air was heated from temperatures of 45° to 122° up to temperatures of 224° to 363°. From 8,035 to 10,070 Th. U. were given to the air per pound of coal used. About 5,200 Th. U. were transmitted to the air per hour per square foot of heating surface.

Combination of a Gas Motor and Air Engine.—In a scheme for distributing power, chiefly by compressed air, for the town of Dresden, Dr. Pröll proposed to work an electric lighting station partly by air motors and partly by gas engines. The ordinary re-heating apparatus for air motors is not very convenient in this case, in consequence of the great variation in the demand for power. Hence Dr. Pröll adopted the plan of combining gas engines with air motors. The gas engine is itself a very efficient and convenient motor for an electric lighting station, because it can be put in action or stopped according to the variation in the demand for power, and there is no waste like that due to keeping boilers in steam ready for use. But in gas engines a very large fraction of the heat developed is necessarily wasted in the water-jacket. Dr. Pröll proposed to abolish the water-jacket and to take the compressed air through the gas-engine jackets to re-heat it on its way to the air motors. In addition, the hot gases rejected from the gas engine were to be used in the jacket of the air motors. Undoubtedly by the combination of the gas engine and the air motor a quite remarkable thermal efficiency could be obtained.

It may be questioned whether it would not be better to take the exhaust of the gas engine directly into the air current. Then the gas engine would work with a heavy back pressure, but the work so lost would be recovered in the air motor. In a paper on compressed air,[1] the author suggested re-heating by the burning of gas in the air current, so that the whole of the heat would be utilised without chimney losses. Some attempts have

[1] 'Transmission of Power by Compressed Air,' *Proc. Inst. C.E.*, vol. xciii.

since been made in this direction in America. Fig. 55 shows a small petroleum burner used in the air main supplying compressed air to rock-drilling machinery.

Meters for Measuring Air supplied to Consumers.—Various types of meters have been used in compressed air systems. Very accurate displacement or positive meters can be constructed, but they are costly. Hence inferential meters, which are virtually air turbines driven by the air current, are more commonly used. Fig. 56 shows an arrangement designed by Mr. Abrahams, of the Birmingham Compressed Air Company, which is stated to have worked with an accuracy within 1 per cent. With a simple fan or turbine driven by the air current the velocity of the meter is not proportional to that of the air current, in consequence of the friction of the meter. If set to be right at a mean velocity, it over-registers with a fast current and under-registers with a slow current. Mr. Abrahams added a kind of pendulum governor, the balls being replaced by hemispherical cups. The governor creates a resistance increasing with the radius of the circle in which the cups revolve, and therefore with the speed of the meter. This extra resistance may be made to balance the tendency to over-register.

Fig. 55.

Cost of Working with Compressed Air.—Air motors can be obtained erected complete for two-thirds of the cost of a steam-engine and boiler. In the very imperfect small rotary motors in Paris, the consumption of air compressed to five atmospheres (75 lbs. per sq. in.) is 750 to 850 c. ft. of air at atmospheric pressure per effective h.p. hour. Old steam-engines converted to air motors use 450 c. ft. of air at atmospheric pressure per effective h.p. hour.

Now the new compound air compressors at Paris compress a cubic metre of air (at atmospheric pressure) to more than six atmospheres for 0·4 centime. That is 1·08d. per 1,000 c. ft. If the selling price is taken at 2d. per 1,000 c. ft., to allow for interest and depreciation on plant, this would correspond to 1·6d. per effective h.p. hour for air used in the inefficient rotary motors, or $\frac{9}{10}d$. per effective h.p. hour for air used in the converted steam-engines of tolerably good efficiency. In the latter case the cost amounts to 11*l.* 5*s.* per effective h.p. per year of 3,000 working hours. Allowing for interest on the cost of motor and wages, the cost per h.p. per annum would be about 13*l.* Better results than this may be expected when air motors are constructed as carefully as air compressors.

The prices charged for air in Paris have not been very

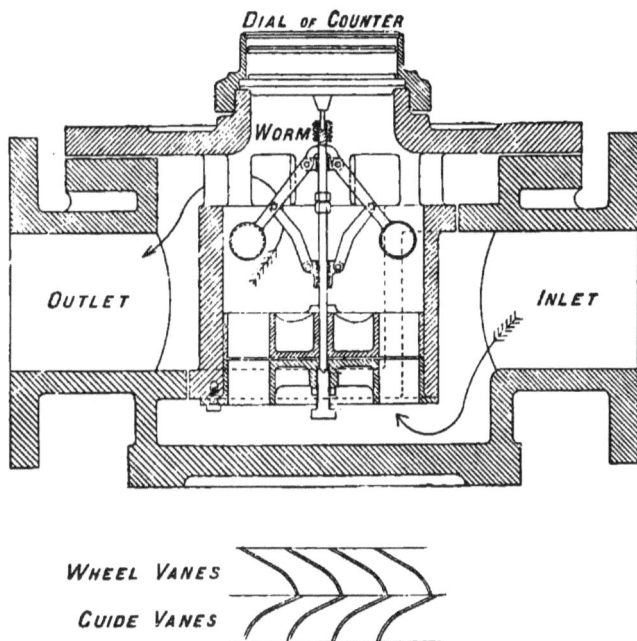

FIG. 56.

authoritatively published. It has been stated that 1·5 centime per cubic metre is charged. This would make the cost of power about double that calculated above. But the Paris prices were settled in the early days, when very extravagant and wasteful compressing plant was in operation.

Distribution of Power by Compressed Air at the Works of the Société Cockerill at Seraing.—The great works at Seraing may be considered the birthplace of modern compressed-air machinery. The compressed-air plant for the Mont Cenis Tunnel works was made at the Cockerill works, having been designed and con-

structed under the direction of Mr. J. Kraft, who is now at the
head of the engineering staff of the works. In conjunction
with M. Sommeiller, Mr. Kraft carried out extensive experiments
on the efficiency of air-compressing machines, in order to obtain
the necessary data as a guide in attacking what was then a new
problem. Compressed air in mines was first and is still extensively used at the Marihaye collieries at Seraing. Further,
since 1854 compressed air has been used in the engine works
of the Société Cockerill for working cranes.[1] With regard to
this last application, Mr. Kraft states that ' it might be expected
that the losses of power incurred in the production and the
utilisation of compressed air would cause it to be rejected as a
motive power. But in many cases it is not so ; for instance, for
a series of cranes, machines working only at intervals. Where
steam is used, enormous losses are caused by condensation in
the pipes, and expansion and condensation can hardly be used
in the engines. Whereas, on the other hand, compressed air
can be produced by high-class engines consuming very little
coal. In this way the loss incurred by employing air may be
compensated for. For a set of cranes like those at the Cockerill
works or at Portsmouth Dockyard, steam cannot compete with
air. The principal rival of compressed air is water, and there
are many cases where water is to be preferred. For cranes
placed in the open air in cold countries the great impediment to
the use of water is frost.' For the installation of a number of
cranes in the open air, along a quay wall, Mr. Kraft thinks
that air is preferable, as in the case of Portsmouth Dockyard.
The compressed-air machinery in the engine works at Seraing
in 1885 (and still in use) consisted of the following machines :

(a) Air-compressing engine with two cylinders. Diam.
of steam cylinders and air cylinders, 13·78 ins. ; stroke, 29·53 ins.
Revolutions per minute, 26.

(b) Air reservoirs, two ; length 36 ft., diameter 6½ ft. Maximum pressure, 5 atmospheres. Diameter of pipes, 2 ins.

(c) 40-ton travelling crane with two double-cylinder air
motors. Diameter of cylinders, 4·53 ins. ; stroke, 7·09 ins. The
air is supplied to the traveller by a flexible pipe which coils on to
or off a drum as the traveller approaches to or recedes from one

[1] 'Notes on Compressed Air and Machinery for Utilising it,' by John
Kraft; *Proc. Inst. Civil Engineers*, vol. lxxxix.

end of the building. The crane has worked very satisfactorily, and was at work this year (1893).

(d) Three 4-ton swivel cranes.

(e) Air motor working hydraulic pumps for wheel-press.

(f) Twelve 1½-ton swivel cranes.

(g) Two 15-ton swivel cranes.

The compressing engine stops automatically when the pressure reaches 75 lbs. per sq. in., and begins working again when the pressure falls. The air cranes differ in no respect from steam cranes so far as their motors are concerned, and all can be worked with a pressure of 45 lbs. per sq. in. Also an overhead traveller with a reservoir of air at 90 lbs. has been erected in the foundry. The air is supplied to the motor through a reducing valve at 40 lbs. per sq. in.

Compressed Air Plant at Portsmouth Dockyard.[1]—In the most modern part of the dockyard the lifting and hauling appliances are chiefly worked by compressed air. There are two sets of compressors, one working to 90, the other to 200, i.h.p. The air is compressed at 60 lbs. pressure per sq. in. into eight wrought-iron receivers of 18,000 c. ft. total capacity. The air mains have a total length of 14,000 feet and vary from 3 inches to 12 inches diameter. There are forty 7-ton capstans, five 20-ton cranes, machinery for working seven caissons, and numerous penstocks, all driven by the compressed air.

At Portsmouth there is both an hydraulic and compressed-air distribution of power, and Mr. Corner compares their relative advantages for such work as that required in dockyards. He concludes that the requirements are best and most economically met by a compressed-air distribution. This is largely due to the advantage of having a considerable store of energy in the receivers. There is also less wear in the air plant, and it is less easily put out of working order.

Compressed Air in Mines.—One of the largest mining plants worked by compressed air is that at the Chapin Mine, Michigan.[2] About three miles from Iron Mountain, at Quinesec Falls, on a head of 52 feet, 1,700 h.p. is obtained by four turbines. Each

[1] I. T. Corner: ' Lifting and Hauling Appliances at Portsmouth Dockyard,' *Proc. Inst. Mech. Eng.*, 1892.

[2] See *The Iron and Steel Institute in America*, p. 378.

of these drives two ' Rand ' compressors. About 2,500,000 c. ft. of air are supplied per day at 60 lbs. per sq. in. gauge pressure. From the compressor plant a 24-inch wrought-iron main, $\frac{1}{4}$ inch thick, extends for three miles, an expansion joint being used at every 480 feet. The air main is connected to the machinery and to 105 power drills at the Chapin Mine, and also to some neighbouring mines. Most of the machinery is arranged so that by closing one valve and opening another a change can be effected from working by air to working by steam.

In his address to the Mechanical Section of the British Association at Cardiff, Mr. Foster Brown mentioned the use of compressed air at a coal-mine in South Wales. There compressed air has the advantage that it assists ventilation and is safe in an explosive atmosphere. There are two tandem compound steam-engines, working with steam at 150 lbs. per sq. in. and developing 1,600 h.p. There is one air cylinder behind each low pressure cylinder, 34 inches diameter and 60 inches stroke.

Compressed Air Plant at the Steel Works, Terni, Central Italy.[1]—The locality of Terni was selected for the steel works chiefly because of the large water power available there. From a point above the Marmore Waterfall, water is conveyed to the works and distributed in pipes to the turbines. The total fall from the Velino to the works is 750 feet. The supply canal is 7,217 yards in length, of which 2,900 yards are in tunnel. The canal carries water sufficient to develop 8,000 effective h.p. Part of the hydraulic power is used directly in driving mills, hoists, &c. Part is used to compress air, which is then distributed to work the great hammer and other machinery. One advantage of the compressed-air system is that there are large air reservoirs, holding a supply of energy ready for use, when in working the heavy machinery there is a sudden and large demand.

The compressors are François and Dubois compressors. There are four sets of compressors coupled to a common shaft carrying a fly-wheel. The compressor cylinders are $31\frac{1}{2}$ inches diameter and $47\frac{1}{4}$ inches stroke. They are driven by water cylinders 14 inches in diameter. The maximum speed is 24 revo-

---

[1] Savage: ' Machinery for the new Steel Works at Terni,' *Proc. Inst. C. E.*, vol. xciii.

lutions per minute. The four sets of compressors work to about 2,000 h.p., delivering 1,871,000 c. ft. of air, at an effective pressure of 75 lbs. per sq. in., in the twenty-four hours. The air compressors are cooled by spray injection, the water injected per stroke being about $\frac{1}{100}$th of the cylinder volume. The temperature of the air leaving the compressors is stated to be 120° to 160° Fahr.

The great 100-ton hammer is worked by a single-acting air cylinder 75½ inches diameter and 16½ feet stroke. The 100-ton and 150-ton cranes are worked each by two double-cylinder air engines with cylinders 7·8 inches in diameter and 12 inches stroke, running at a maximum speed of 200 revolutions per minute.

System of Transmitting Motive Power by Vacuum.—An interesting plant for distributing motive power by vacuum was established in the Rue Beaubourg, in Paris, by MM. Petit and Boudenoot.[1] The general object in view was the distribution of power to small industries. From 1874 M. Petit had the idea of transmitting power by vacuum. In 1882 an association was formed and machinery erected. Conduits were laid communicating with the houses of consumers, who paid a rental based on the number of rotations of their machines ascertained by a counter. The users of power were interested in the success of the scheme by participation in the profits. The working hours are from 7 A.M. till noon, and from 1 P.M. till 8. A steam-engine and exhausting pump of 70 to 80 h.p. was first erected, the mains extending 300 to 400 yards. Now there are three steam-engines, developing altogether 300 h.p., and the mains extend 850 yards. There are about 150 small motors on the mains. Part of the power is rented to an electrical company. This power is supplied by a fourth engine of 100 h. p. M. Boudenoot gives the preference to a vacuum system because the cost of machinery is less than for a compressed air system. The mains are always dry and do not require draining-boxes. Lastly, the efficiency of a vacuum system is, in his opinion, greater. M. Boudenoot takes the efficiency of the exhausting pump at 0·93, the mechanical efficiency of the

---

[1] 'Transmission de la Force motrice par l'Air raréfié,' par M. Max de Nansouty, *Génie Civil*, 1886. 'Distribution de la Force motrice à Domicile au moyen de l'Air raréfié,' par M. Boudenoot ; *Mém. de la Société des Ingénieurs Civils*, 1885 and 1889.

vacuum motors at 0·60, and the efficiency of the expanding air in the motor at 0·85. The efficiency of transmission he takes at 0·95. The resultant efficiency is then

$$0·93 \times 0·60 \times 0·85 \times 0·95 = 0·45.$$

This, for small motors, is a good result. The exhausting cylinders make 20 to 50 revolutions per minute, and maintain a vacuum of 0·67 to 0·80 atmosphere. These cylinders have spray injection. The motors are constructed to supply 360, 510 and 900 foot lbs. per second. There is a vacuum reservoir, 50 inches in diameter and 140 inches in height, attached to each motor. The vacuum mains are 10, 8, 6 and 4 inches in diameter ; the house-service pipes are of lead. The vacuum system is undoubtedly very convenient and efficient for a domestic system of this kind. For transmission of large amounts of power to greater distances it is not so well adapted, chiefly on account of the size of the mains, pumps and motors which would be necessary.

The Paris System of Distribution of Power by Compressed Air.—The Paris power distribution is at present the largest in Europe, but it developed out of very small beginnings. About 1870, MM. Popp and Resch established, first in Vienna and then in Paris, a system for regulating clocks by impulses of compressed air. At first, in Paris, there was a central station in the Rue Argenteuil, with two small compressors delivering air into a receiver at 2 to 3 atmospheres pressure. In a second receiver air was maintained at a constant pressure of $1\frac{3}{4}$ atmosphere. Two clocks (one in reserve) actuated a distribution valve, allowing air to pass into the mains for 20 seconds in each minute. By means of small pipes, laid chiefly in the sewers, the air impulses were conveyed to the clocks which were to be regulated. At the clock a small bellows lifted a rod at each impulse and moved the escapement of the clock. The air mains were generally $\frac{5}{8}$ to $\frac{7}{8}$ inch in diameter, and the service pipes into the houses $\frac{1}{4}$ to $\frac{3}{8}$ inch in diameter. These pipes were extended over many miles, and in 1889 there were in Paris 8,000 clocks regulated by the pneumatic arrangements. The system proved so successful that a new central station was erected in the Rue St.-Fargeau, in Belleville. Down to 1887, two small Farcot engines and a beam engine sufficed for the work.

Gradually there arose a demand to use the compressed air for small motors. An extension of the station was then made, and a second installation erected in the Rue St.-Fargeau. This consisted of six Davey Paxman compound engines, each driving two compressing cylinders. The engines developed 2,000 h.p. The compressors were made in Switzerland, on the Blanchard system. Soon this plant became insufficient. A third installation was erected in the Rue St.-Fargeau in 1889.[1] This consisted of five compound engines and compressors, built by the Société Cockerill at Seraing, the compressors being on the Dubois-François system. These engines developed 2,000 h.p. Finally, another central station, at the Quai de la Gare, was

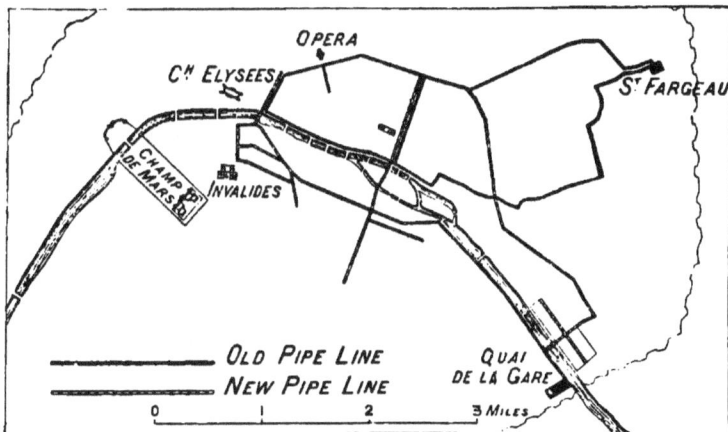

FIG. 57.

erected, with engines of 8,000 h.p., and room for extension to 24,000 h.p. The engines are triple vertical engines, with Riedler compound compressors. The air is compressed so much more cheaply at the new station, that the old station in the Rue St.-Fargeau is no longer worked.[2]

At the St.-Fargeau station neither the engines nor the compressors were of the best type. Their effect was to compress

---

[1] See the sketch-map of the air-mains in Paris, fig. 57.

[2] For numerous details and data relating to the Paris system, see *Neue Erfahrung ueber die Kraftversorgung von Paris durch Druckluft*, von A. Riedler, Berlin, 1891; *La Distribution de la Force par l'Air comprimé dans Paris*, par Prof. Riedler, Paris, 1891. See also 'Note sur le Transport de l'Energie par l'Air comprimé,' par M. Solignac, *Congrès International de Mécanique Appliquée*, Paris, 1893.

265 c. ft. of air (20·25 lbs.) at atmospheric pressure to 6 atmospheres per i.h.p. per hour. The cooling in the compressor cylinders was ineffective. The air from the six compressors was delivered into eight cylindrical receivers of 1,150 c. ft. capacity each. The air was then distributed by cast-iron mains, 11·8 inches in diameter; these had joints with india-rubber rings, forming a kind of stuffing-box and permitting expansion at every pipe length without leakage. The mains were laid partly under roadways and partly in sewer subways. They were supplied at intervals with automatic draining-boxes. It was remarkable that the demand for power from this station so rapidly grew up to its full capacity.

The air motors used are generally of a very simple kind. For small powers a simple rotary engine is used; for larger powers steam-engines are employed, worked with air instead of steam. Where the compressed air enters a building it generally passes through a screen, which removes solid impurities. Then there is a stop valve and a meter for measuring the air. Next there is often a reducing valve, by which the pressure is reduced to $4\frac{1}{2}$ atmospheres. In most cases there is a re-heater, often a simple double-walled box of cast iron, in which the air circulates and is heated by a coke fire. For a 10-h.p. motor this re-heater is about 21 inches in diameter and 33 inches high. The amount of coke used is not considerable, about $\frac{1}{4}$ lb. per h.p. per hour. The air is raised in temperature to 300° Fahr. The air motors are very convenient. They can be started at any moment, they are free from inconvenience from leakage, heat or smell, and they require a minimum of attendance. Often the exhaust can be used to cool and ventilate the working rooms. The air motors are used for various purposes. At some of the theatres and restaurants they drive dynamos for electric lighting. At some of the newspaper offices there are motors of 50 and 100 h.p. driving printing-machines. In workshops there are motors driving lathes, saws, polishing, grinding, sewing, and other machines. At the Bourse de Commerce the compressed air drives dynamos for electric lighting, and also is used to produce cold in large refrigerating stores. In many of the restaurants air is used for cooling purposes. It is also used to work cranes and lifts directly, a water-cushion being used between the working cylinder and the lift.

In the first compressors at the Rue St.-Fargeau about 265 c. ft. (20·25 lbs.) of air were compressed to 6 atmospheres, per i.h.p. hour. The second installation had somewhat better compressors. These c o m p r e s s e d about 300 c. ft. (22·92 lbs.) of air per i.h.p. hour. Later, permission was given to Professor Riedler to convert one of the Cockerill compressors into a compound compressor. After the change, 370 c. ft. (28·27 lbs.) of air were compressed to 6 atmospheres per i.h.p. hour. In the large new compressors, at the Quai de la Gare, 438 c. ft. (33·47 lbs.) of air are compressed per i.h.p. hour. The general result of the working of the new station is that a cubic metre of air is compressed to 7 atmospheres for 0·4 centime,[1] or about four-tenths of the cost of compression at the older station.

In the new station, fig. 58, the steam-engines are vertical triple engines, work

[1] That is, 883 c. ft. or 67½ lbs. of air compressed for one penny.

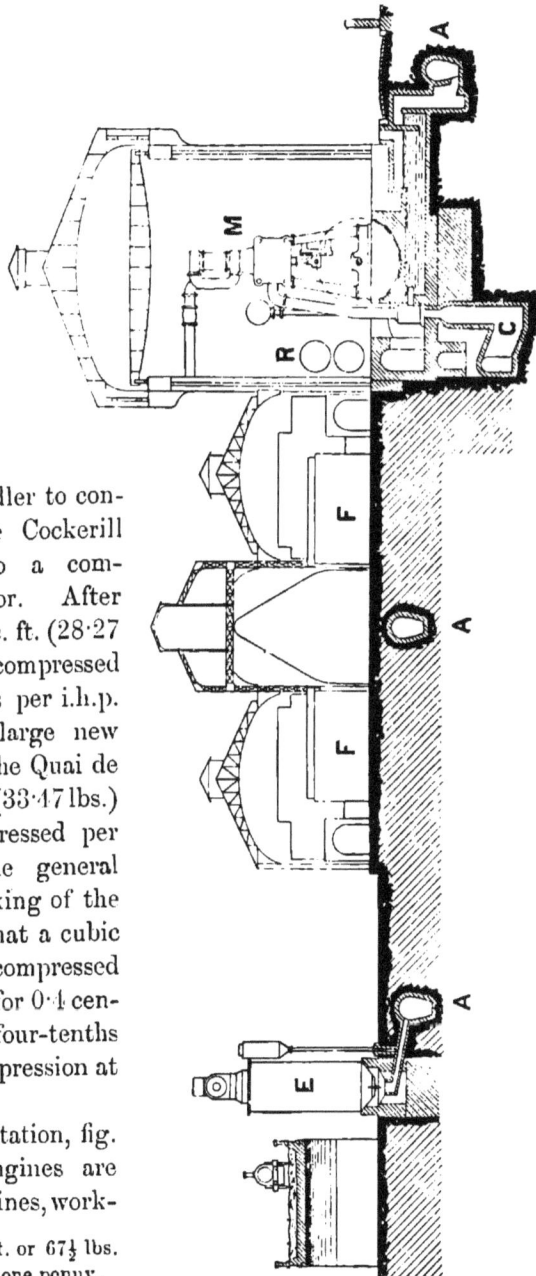

FIG. 58.

ing compound compressors with two low and one high pressure compressing cylinders. Each engine is of 2,000 h.p.[1] Four have been erected, three being regularly worked and one kept in reserve. The new station has been placed on the banks of the Seine, where coal and condensing water can be obtained cheaply. Steam of 180 lbs. per sq. in. pressure is used, and the makers guaranteed that the engines would work with 1·5+ lbs. of coal per i.h.p. hour. The new air main, about 7 kilometres in length, is 20 inches in diameter.

Compressed Air System at Offenbach, near Frankfort-on-Main. A compressed-air distribution of the most improved type has been constructed at Offenbach, by the firm of Riedinger, of Augsburg. The compressing station has a horizontal compound steam-engine with crank-shaft and fly-wheel. The air compressors are compound, of Riedler's design. The air is compressed to 2 atmospheres in the first cylinder, and then to 6 atmospheres in the second. The heat is abstracted in an intercooler between the high and low pressure cylinders. The steam cylinders are 22 and 31 inches diameter and 40 inches stroke. The air cylinders are 16 inches and 24 inches diameter and 40 inches stroke. Each air cylinder is worked direct from the back of the corresponding steam cylinder. The engine is stated to work with 15½ lbs. of steam per i.h.p. hour, and the efficiency of the compressor is 87 per cent. The engine runs at 75 revolutions, or 490 feet of piston speed per minute. The air is distributed through 23,000 feet of cast-iron mains with india-rubber ring joints. The main, when tested with a pressure of 6¾ atmospheres, kept up for 70 hours, showed a leakage of only 1·6 c. ft. per mile per hour.

Compressed Air Tramways at Berne.[2]—The difficulty of replacing horse traction by mechanical traction on tramways in towns is very considerable. Steam-engine traction has obvious inconveniences. Wire-rope traction has been remarkably successful in certain cases, but it can only be adopted where a large traffic is to be carried. Electric traction with overhead conductors (the trolley system) has been adopted very extensively

[1] In fig. 58 E indicates the position of the water-purifying apparatus; F F the boilers; M the 2,000 h.p. compressors; R the air receivers; C a channel for water supply; A A A sewers for removing water.

[2] See a paper by Mr. Preller, *Engineering*, February 24, 1893.

in America, but it does not appear to be suitable for European towns. Traction by compressed air is possible, and is an interesting case of distribution of power from a central station.

The first system of compressed-air traction which proved successful is that of Mekarski, adopted at Nantes, in Paris, and at Berne. The Berne tramway has been working since 1890, and has gradients of 5 to 6 per cent. The air is compressed to 350 to 450 lbs. per sq. in. in reservoirs of about 75 c. ft. capacity, carried by each tramcar. Between the reservoir and the motor is a small tank of super-heated water at a temperature of about 350° F., supplied from a boiler working at 100 to 150 lbs. per sq. in. The air circulates through the water, and a mixture of air and vaporised water passes to the motors, which are coupled to the car-axles. The heat furnished makes it possible to work with a considerable ratio of expansion. The steam yields its latent heat to the air before exhaust.

At Berne a water-power station has been established on the Aare, at a point where there is a supply of 700 c. ft. per second on a fall of 6½ feet. There are three 120 h.p. pressure turbines, of which two are used for electric lighting, and one for compressing air. The power of this turbine is rented to the tramway company at 4l. per h.p. per annum. There are four compound compressors running at 80 revolutions per minute, of which three are ordinarily sufficient, the fourth being in reserve. Each compressor has two single-acting cylinders. The low-pressure cylinder is 11·8 ins. diameter, and compresses the air to 75 lbs. per sq. in. The high pressure is 5·3 ins. diameter, and compresses the air to 470 lbs. per sq. in. The air is cooled by a water-jacket on the high-pressure cylinder, and by spray injection in the low-pressure cylinder. Each compressor delivers about 350 lbs. of air per hour. The air passes through a separator to drain off the water, and then into two reservoirs of 44 c. ft. capacity. A wrought-iron main 1·3 inch in diameter conveys the air to the charging station for the cars. At the charging station there are six reservoirs of 44 c. ft. capacity each. There are also three steam boilers, two with 35½ and one with 97 sq. ft. of heating surface, worked at 100 lbs. per sq. in. pressure. These serve to charge the heating chambers on the cars with heated water. After each journey the cars are re-charged with air and heated water by pipes and valves connected

with the air reservoirs and air main, and the steam boilers. The
tramcar has two coupled axles with a wheel base of 5·2 feet, and
wheels 27½ ins. diameter. The air motors have cylinders 5·1
ins. diameter, and 8·6 ins. stroke. The total weight of the car
when full is 9½ tons. The authorised speed is 11 kilometres
(6·8 miles) per hour. The maximum initial pressure in the
cylinders is 176 lbs. per sq. in. On easy gradients it is less.
Each car carries twelve air-storage cylinders of 18 ins. diameter.
Aggregate capacity, 75 c. ft. Fully charged they contain 177
lbs. of air at 30 atmospheres. This is sufficient for a journey of
6 kilometres (nearly 4 miles). The heating chamber has a
capacity of 3·5 c. ft.

The cars are charged first from the station air reservoirs, and
then from the pumping main, till the pressure reaches the working
limit. Then the car is disconnected, and the compressor recharges
the reservoirs. This change of action is effected without stopping
the compressor. Reloading a car takes about 15 minutes. The
cars use from 28 to 42 lbs. of air per car mile, or, on the
average, 35 lbs. The line is nearly two miles in length. There
is a 10-minute service. The double journey over the line takes
40 minutes. The working expenses amount to 8·9d. per car
mile, and the receipts to 10·7d. per car mile, so that there is
a return of 4½ per cent. on the capital expended. The system
at Berne has proved convenient and satisfactory. The cars run
smoothly, and there is no annoyance of smoke or smell. There
is a somewhat narrow limit, however, to the distance the cars
will run without recharging.

Compressed Air Tramway: Hughes and Lancaster's System.
This system has not yet been practically adopted, but an
experimental car has been built at Chester, and it has features
which are interesting. The car carries air reservoirs, as in the
Mekarski system, but the air is used at moderate pressure. The
distinctive feature of the system is that an air main is laid along
the tram-line, and arrangements are adopted for recharging
the car, almost automatically, from the air main at convenient
distances along the line.

The air is used at 150 lbs. per square inch pressure, and the
receivers on the tramcar have a capacity of 100 cubic feet. An air
main of 3 or 4 inches in diameter conveys the air to charging
valves underground at the side of the tram-rail. As the car

approaches a charging valve, the conductor, by the movement
of a lever, lowers a plough, which lifts the hinged lid over the
charging valve. The charging valve has four radiating discharge
pipes. A valve on the car has two receiving pipes. By a cam
action, one of the receiving pipes engages with one of the discharge pipes on the air main as the car moves over the charging
box. As the car moves on the valve opens, then closing again,
while the radiating discharge pipes rotate through 90°. As the
receiving pipe disengages, both it and the discharge pipes are
left in position to repeat the charging operation. The car may
be stopped for a few seconds in the position for charging, and
as it moves on again the cover of the charging box falls, leaving
the road surface unbroken.

In the experimental car the air motor was a Rigg engine;
but on the trials this was in an imperfect condition, and air was
lost by leakage. The car, starting with a pressure of about 150 lbs.
per square inch, ran distances of 600 to 1,500 feet with a full load
of passengers, and on gradients varying from level to 1 in 45.
In a special test it ran with one charge of air a distance of
1,230 feet up a steep hill, with gradients of 1 in 130 to 1 in 24.
From the data obtained in tests made by the author, it appeared
that with a reservoir capacity of 100 cubic feet the car would run,
with one charge of air at 170 lbs., distances varying from 7,000
feet on the level to 750 feet on a gradient of 1 in 25. With a
more satisfactory air motor these distances would be exceeded.
This system has the very great advantage that there is no limitation of the distance which the cars will work. It is almost as
convenient as an electric system in which the supply of energy
is continuous, and is free from the difficulties and expense in
electric systems in which the conductors are underground.

In running down inclines of sufficient steepness, the motor
on the Hughes and Lancaster car may be reversed, so as to act
as an air compressor, and it then recharges the air-receivers.
The surplus work is thus utilised, and the motor acts as a brake.

The Shone System of Pumping Sewage by Compressed Air.—
A very interesting system of pumping sewage at many sub-
stations in a flat town district has been invented by Mr. I. Shone.
It has been adopted with success at Henley, at Rangoon, at the
Chicago Exhibition, and at other places. It is strictly a system
of distributing power for pumping purposes from a central station,

and is a good illustration of the convenience of compressed-air transmission for certain purposes.

In order that sewers may be self-cleansing, they must be as small as is consistent with discharging the required quantity, and must be laid at a gradient which secures a sufficient velocity of flow. In an ordinary system of gravitating sewers, the lowest point available is adopted for the outfall, and the sewers are laid with slopes required to give the necessary velocity of flow. But to secure this condition the sewers would in many cases descend below the outfall level. Then the sewers must be constructed at great depths, and pumping at the outfall must be resorted to. The cost of constructing sewers at a great depth and the expense of pumping are so serious, that the engineer is tempted to adopt larger sewers with smaller slope, and to trust to flushing and other arrangements to remove the deposits which form in such conditions. But large flat sewers entail many evils.

With ordinary methods of pumping it is usually necessary to have a single pumping station, to secure economy of superintendence. The fundamental principle of the Shone system is to have a number of pumping sub-stations conveniently distributed, each dealing with the sewage of a definite low area. These sub-stations are worked automatically, without superintendence, by compressed air distributed from a single central station. A town which could not be drained as a whole, by sewers with proper gradients, is divided into smaller districts, each of which can be sewered with self-cleansing sewers draining to one low point in the district, at a level generally not more than 15 feet below the surface. At each of these low points there is an ejector station, worked by compressed air from the central station. The ejectors lift the sewage intermittently into a second system of outfall sewers, according to the rate at which it collects in the ejectors. The outfall sewers are closed pipes without gratings or man-holes, and may be either laid with a uniform gradient, or treated as pumping mains and laid without reference to gradient, the sewage being forced through them by the ejector. For high lifts, in cases where the sewage has to be pumped through a rising main to be distributed over land, two or more ejectors may be used, each dealing with a part of the lift suitable for the air pressure employed.

The ejector (fig. 59) is a simple form of pump for crude

unstrained sewage. It is a closed cast-iron tank, with inlet pipe
and inlet valve and outlet pipe and outlet valve. There is also
an air valve worked by a float. The sewage flows in by gravitation till the ejector is full. At that moment a bell-shaped
float A rises and opens the air valve E. The compressed air entering the ejector closes the inlet valve C and opens the outlet valve
D, driving the sewage into the outfall sewer. As the sewage

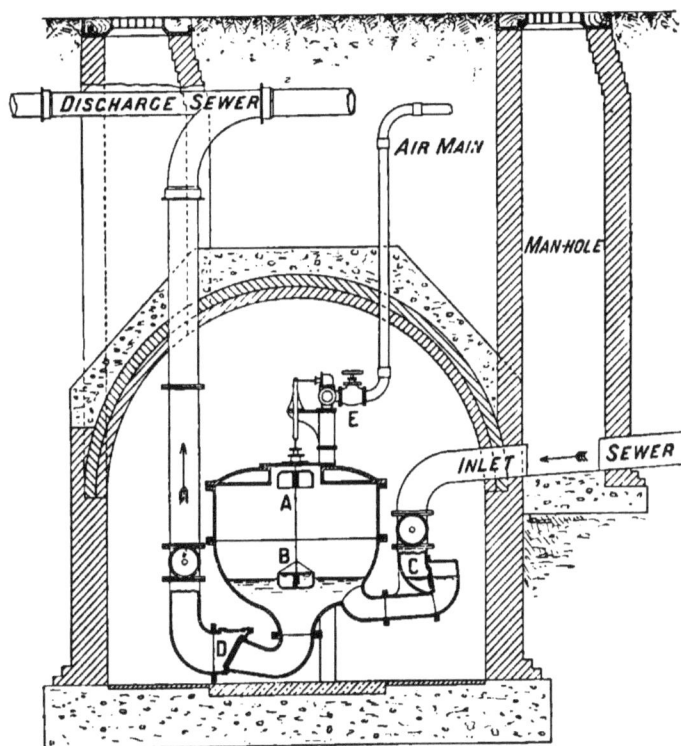

Fig. 59.

falls, it at last leaves unsupported a cup B which closes the air
valve and allows the ejector to refill. The whole operation of
ejection takes only about half a minute. Air mains must be
laid from the central station to the ejector stations. These
are ordinary cast-iron pipes $2\frac{1}{2}$ to 4 inches in diameter.

The author tested the efficiency of the ejector system in
1888. Three ejectors, of 500 gallons capacity each, were
arranged to pump on lifts of 12 to 21 feet. The air was com-

P

pressed by a gas engine. The discharge from the ejectors was measured in a carefully gauged tank.

TRIAL I.—Total lift from bottom of ejector to point of discharge, 24·26 feet. Useful lift from centre of ejector inlet pipe to point of discharge, 21·26 feet. Speed of working, 20 discharges per hour from each of the three ejectors. Air pressure, 11 lbs. per sq. in. Rate of pumping, 331 gallons per minute.

		H.P.
Brake h.p. of gas engine		4·976
Work due to useful lift	2·133	
„ „ „ fall into ejector	·302	
Friction of compressor	·653	
Loss in heating air	1·516	
Other losses	·372	
		4·976

TRIAL II.—Total lift from bottom of ejector to point of discharge, 12·33 feet. Useful lift from centre of ejector inlet pipe to point of discharge, 9·33 feet. Speed of working, 26 discharges per hour from each of the three ejectors. Rate of pumping, 432 gallons per minute. Air pressure, 6 lbs. per sq. in.

		H.P.
Brake h.p. of gas engine		3·242
Work due to useful lift	1·222	
„ „ „ fall into ejector	·392	
Friction of compressor	·653	
Loss in heating air	·663	
Other losses	·312	
		3·242

It will be seen that the resultant efficiency of compressor, mains, and ejectors is 49 per cent. in both trials, reckoning on the total lift. It is 42 per cent. in the first trial and 38 per cent. in the second, reckoning on the useful lift, which is exclusive of the drop from the inlet pipe into the ejectors. In any system of pumping some loss is unavoidable from a drop into the collecting tank from which the sewage is pumped, and the loss is proportionately greater the less the lift. This is a very satisfactory efficiency on such low lifts. The working of the injectors was entirely automatic, and they required no attention.

CHAPTER X

CALCULATION OF A COMPRESSED AIR TRANSMISSION WHEN THE SUBSIDIARY LOSSES OF ENERGY ARE TAKEN INTO ACCOUNT [1]

Loss of Pressure in the Air Mains.—Hitherto information as to the resistance of air mains has been scanty. The best experiments were those made by Mr. Stockalper on the pipes of the boring machines at the St. Gothard Tunnel. The new experiments carried out by Professor Riedler and Professor Gutermuth, [2] on the air mains in Paris, are therefore of great value. The Paris mains are larger than any hitherto tried, and by coupling up different mains at night, a length of $10\frac{1}{4}$ miles could be experimented on.

The main for the older Paris compressing station consists of cast-iron pipes $11\frac{3}{4}$ inches or 0·98 foot in diameter. It was laid partly in the sewers, which involved the use of a good many bends. Part of the main in the Rue de Belleville is known to be leaky, and there are numerous draining-boxes, siphons, and stop-valves which cause resistance. In a more perfectly arranged main no doubt the loss of pressure would probably be somewhat less than in this old one in Paris. In the following investigation Professor Riedler's results are used, but the reductions from them and the conclusions deduced are the Author's. The formula for the flow of air adopted is that given in the Author's paper on 'The Motion of Light Carriers in Pneumatic Tubes.' [3]

---

[1] This chapter is mainly a reprint, by permission of the Council of the Institution of Civil Engineers, of a paper by the Author in the *Minutes of Proceedings*, vol. cv.

[2] *Neue Erfahrungen über die Kraftversorgung von Paris durch Druckluft*, Berlin, 1891.

[3] *Minutes of Proceedings Inst. C. E.*, vol. xliii., 1876. Also article on 'Hydromechanics,' *Encyclopædia Britannica*, ninth edition. Hydraulics, § 84.

Formula for the Flow of Air in Long Pipes.—Let d be the diameter and l the length of a pipe in feet; v the velocity of the fluid in feet per second, and h the head lost measured in feet of fluid under the given conditions. Then, if the fluid is incompressible,

$$\zeta \frac{v^2}{2g} = \frac{d}{l} \cdot \frac{h}{l}$$

$$h = \zeta \frac{v^2}{2g} \cdot \frac{4l}{d}.$$

This formula may be used for the flow of air in pipes, and indeed has been so used by Mr. Stockalper and others, when the variations of pressure and density are small, so that mean values can be taken and the variations neglected. Professor Riedler uses this equation also, but with the artifice of dividing a long main into portions calculated separately — a method very cumbrous and not very accurate.

When air flows along a pipe there is necessarily a fall of pressure due to the resistance of the pipe, and consequently the volume and velocity of the air increase going along the pipe in the direction of motion. The effect of the resistance is to create eddying motions, which, as they subside, give back to the air the heat equivalent of the work expended in producing them. The result is that, apart from conduction to external bodies, the flow is isothermal. Generally, in compressed air systems, the air is delivered into the mains at a temperature above that of the surrounding earth. The excess of heat is parted with by conduction, and the temperature falls to that of the ground, but no lower, for there can then be no further loss of heat by conduction.

Let P = the absolute pressure of the air in lbs. per square foot.

T = the absolute temperature.

G = the weight of the air per cubic foot in lbs.

V = the volume of the air per lb. in cubic feet.

Then,
$$PV = \frac{P}{G} = cT \qquad . \qquad . \qquad (1)$$

where $c = 53\cdot18$ for air. Taking the temperature of $60°$ Fahrenheit, so that $T = 461 + 60 = 521$—

$$cT = 53\cdot18\ T = 27{,}710. \qquad . \qquad . \qquad (2)$$

If air is flowing steadily in a pipe, the same weight of air flows across every transverse section per second. Hence, if w is the weight, in lbs., of air flowing per second, Ω the area of a cross-section, at which the velocity is u—

$$w = G\Omega u = \text{constant} \qquad . \qquad . \qquad (3)$$

combining (1) and (3)

$$\Omega u P = cTW \qquad . \qquad . \qquad . \qquad . \qquad (3a)$$

Fig. 60 represents a short length dl of an air main, between transverse sections A_0, A_1. Let d be the diameter, Ω the cross-section, m the hydraulic mean radius of the pipe. Let P and u be the pressure and velocity at A_0, $P + dP$ and $u + du$ the same quantities at A_1. Let W be the weight of air flowing through the pipe per second. The units are feet and lbs.

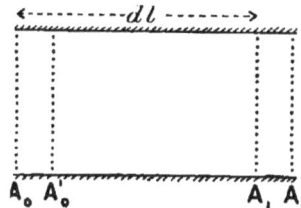

FIG. 60.

If in a short time dt the mass A_0A_1 comes to $A'_0A'_1$, then $A_0A'_0 = u\,dt$ and $A_1A'_1 = (u + du)\,dt$.

By analogy with liquids the head lost in friction measured in feet of fluid is—

$$\zeta \frac{u^2}{2g} \frac{dl}{m}.$$

Let $H = \dfrac{u^2}{2g}$, then the head lost is

$$\zeta H \frac{dl}{m},$$

and since W dt is the flow through the space considered in the time dt, the work expended in friction is—

$$-\zeta \frac{H}{m} W\, dl\, dt.$$

The change of kinetic energy in the time dt is the difference of the kinetic energy of $A_1A'_1$ and A_0A_0', that is—

$$\frac{W\, dt}{2g} \left\{ (u + du)^2 - u^2 \right\}$$

$$= \frac{W}{g} u\, du\, dt = W\, dH\, dt.$$

DISTRIBUTION OF POWER

The work of expansion of $\Omega u\,dt$ cubic feet of air to $\Omega\,(u + du)$ dt, at a pressure initially P, is $\Omega\text{P}\,du\,dt$. But from (3a)

$$u = \frac{c\text{TW}}{\Omega\text{P}}.$$

$$\frac{du}{d\text{P}} = -\frac{c\text{TW}}{\Omega\text{P}^2},$$

and the work done by expansion is—

$$-\frac{c\text{TW}}{\text{P}}\,d\text{P}\,dt.$$

The work done by gravity is zero if the pipe is horizontal, and in most cases may be neglected without great error. The work of the pressures on the sections A_0A_1 is

$$\text{P}\Omega u\,dt - (\text{P} + d\text{P})\,\Omega\,(u + du)\,dt$$
$$= -(\text{P}du + ud\text{P})\,\Omega dt.$$

But if the temperature is constant—

$$\text{P}u = \text{constant}$$
$$\text{P}du + ud\text{P} = 0$$

and the work of the pressures is zero.

Adding together the quantities of work and equating them to the change of kinetic energy—

$$\text{W}\,d\text{H}\,dt = -\frac{c\text{TW}}{\text{P}}\,d\text{P}\,dt - \zeta\frac{\text{H}}{m}\text{W}\,dl\,dt$$

$$d\text{H} + \frac{c\text{T}}{\text{P}}\,d\text{P} + \zeta\frac{\text{H}}{m}\,dl = 0$$

$$\frac{d\text{H}}{\text{H}} + \frac{c\text{T}}{\text{HP}}\,d\text{P} + \zeta\frac{dl}{m} = 0 \quad . \qquad . \qquad (4)$$

But—

$$u = \frac{c\text{TW}}{\Omega\text{P}}$$

$$\text{H} = \frac{u^2}{2g} = \frac{c^2\text{T}^2\text{W}^2}{2g\Omega^2\text{P}^2}$$

$$\frac{d\text{H}}{\text{H}} + \frac{2g\Omega^2\text{P}}{c\text{TW}^2}\,d\text{P} + \zeta\frac{dl}{m} = 0.$$

For pipes of uniform section Ω and m are constant, for steady motion w is constant, and for isothermal flow T is constant. Integrating—

$$\log H + \frac{g\Omega^2 P^2}{w^2 cT} + \zeta \frac{l}{m} = \text{constant} \quad . \quad (5)$$

For $l=0$, let $H=H_1$ and $P=P_1$

,, $l=L$, let $H=H_2$ and $P=P_2$

$$\log \frac{H_2}{H_1} + \frac{g\Omega^2}{w^2 cT}(P_2{}^2 - P_1{}^2) + \zeta \frac{L}{m} = 0 \quad . \quad (6)$$

where P_1 is the greater and P_2 the less pressure. L is the length of transmission. By replacing H_1, H_2 and w—

$$\log \frac{P_1}{P_2} + \frac{gcT}{u_1{}^2 P_1{}^2}(P_2{}^2 - P_1{}^2) + \zeta \frac{L}{m} = 0 \quad . \quad (6a)$$

Hence the initial velocity in the pipe is—

$$u_1 = \sqrt{\left\{ \frac{gcT(P_1{}^2 - P_2{}^2)}{P_1{}^2\left(\zeta \frac{L}{m} + \log \frac{P_1}{P_2}\right)} \right\}}$$

and when L is great, $\log \frac{P_1}{P_2}$ is comparatively small compared with the other term in the bracket. Then—

$$u_1 = \sqrt{\left\{ \frac{gcTm}{\zeta L} \frac{P_1{}^2 - P_2{}^2}{P_1{}^2} \right\}} \quad (7)$$

For pipes of circular section $m = \frac{d}{4}$, where d is the diameter in feet. Let $cT = 27{,}710$, and let p_1, p_2, be the pressures in lbs. per square inch. Then—

$$u_1 = \sqrt{\left\{ 222{,}000 \frac{d}{\zeta L} \frac{p_1{}^2 - p_2{}^2}{p_1{}^2} \right\}} \quad . \quad . \quad (7a)$$

This equation is easily used. In some cases the approximate equation—

$$u_1 = \left(1{\cdot}1319 - 0{\cdot}7264 \frac{p_1}{p_0}\right) \sqrt{\left(222{,}000 \frac{d}{\zeta L}\right)} \quad (8)$$

may be more convenient.

If the terminal pressure p_2 is required in terms of the initial pressure p_1, then—

$$p_2 = p_1 \sqrt{\left\{1 - \frac{\zeta u_1{}^2 L}{222,900 d}\right\}} \qquad (9)$$

If from a series of experiments ζ is to be found—

$$\zeta = 222,900 \; \frac{d}{u_1{}^2 L} \; \frac{p_1{}^2 - p_2{}^2}{p_1{}^2} \qquad (10)$$

Variation of Pressure and Velocity in a Long Main.—In order to have some idea of the law of variation of pressure and velocity in long mains, two cases have been calculated. Taking a main of 12 inches diameter, and assuming $\zeta = 0.003$, the following results are obtained

AT DISTANCES IN MILES FROM ORIGIN OF MAIN

—	0	1	2	3	4	5	6	7	8	9	10
CASE I.— Pressure (absolute) in lbs. per sq. in.	115	112·3	109·8	107·0	104·3	101·4	98·5	96·1	92·3	89·1	85·7
Velocity in main in ft. per sec.	25	25·6	26·2	26·9	27·6	28·4	29·2	29·9	31·2	32·3	33·6
CASE II.— Pressure (absolute) in lbs. per sq. in.	115	104·3	92·3	78·6	61·8	38·5	0				
Velocity in main in ft. per sec.	50	55·1	62·3	73·2	93·1	149·4	∞				

It will be seen that with an initial velocity of 25 feet per second the pressures decrease comparatively slowly and the velocities increase also somewhat slowly. Even at 10 miles the pressure is still considerable and the velocity moderate. With an initial velocity of 50 feet per second, the variation of pressure and velocity is much greater in long mains. Beyond 5 miles the pressure becomes too small to be practically available and the velocity enormous.

These results are plotted in curves in fig. 61.

Before proceeeding to calculate the value of ζ from the experiments of Professor Riedler and Professor Gutermuth on the Paris mains, it is necessary to examine the corrections to be made for leakage and the special resistances of the draining boxes.

Loss of Air by Leakage from the Main.—Special experiments were made by Professor Riedler and Professor Gutermuth to

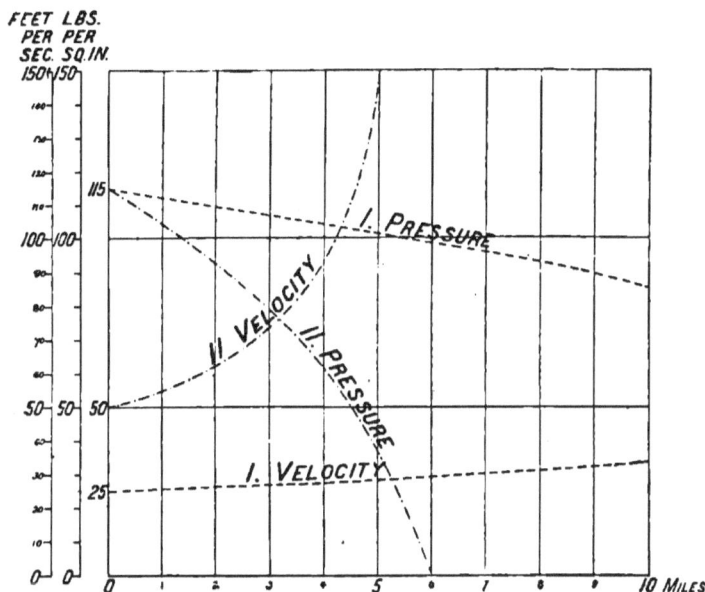

FIG. 61.

determine the leakage loss in portions of the old Paris main. These gave the following results:—

	I	II	III	IV	V	VI
	Southern section. Central Works to Pl. de la Concorde	Entire pipe line	Central Station to Pl. de la Concorde	Entire pipe line	Northern section. Central Station to Rue de Belleville	Rue des Pyrenées
Length in miles .	5·69	10·69	5·69	10·69	0·87	0·34
Initial pressure in atmospheres	6·5	6·9	7·0	6·7	6·0	6·1
Terminal pressure	6·0	5·9	6·43	5·82	5·0	3·7
Loss of pressure per hour in atmospheres	1·5	1·5	0·57	1·32	0·6	0·56
Loss of air in cubic feet per hour, reckoned at atmosphere pressure	34300	67000	13000	59000	2100	782
Loss of air in cubic feet per mile per hour	6030	6270	2285	5520	2414	2300

It is stated that during trials I, II, and IV, considerable leakage was known to exist at the Central Station and the Rue de Belleville. In all the trials the air consumed by the clock system and some small motors not stopped is included. Reduced, as shown in the last line, the results are consistent, if it is assumed that the leakage was really greater in the sections included in I, II, and IV than in the other sections.

Now, with an initial gauge-pressure of 6 atmospheres and an initial velocity of 30 feet per second, the main would deliver about 600,000 cubic feet of air per hour, reckoned at atmospheric pressure. Then the percentage of loss by leakage per mile per hour would be as follows :—

——	I	II	III	IV	V	VI
Per cent. of air lost by leakage per mile per hour	1·00	1·05	0·38	0·91	0·40	0·38

Professor Riedler believes that in newer and better laid mains the leakage is considerably reduced. In any case, it appears that the loss is small and practically negligible except in very long transmissions, at least when the main is delivering a full supply. It should be remembered, however, that the leakage is proportionately greatest in those hours when the pressure has to be maintained in the mains, but the demand for power is small. Where the demand for power is very variable this loss might become an appreciable factor. Looking to the uncertainty as to the amount and law of variation of leakage in different cases, it will be neglected in this investigation, taking the experiments above referred to as a proof that it is not a serious quantity even in the Paris installation, and that in newer mains laid with the experience now gained it would be still less important.

Correction for Loss of Pressure at Draining Tanks.—On some of the mains experimented on by Professor Riedler there were draining tanks at which there was a sensible loss of pressure. The resistance at such a receiver must be analogous to the loss at a sudden enlargement of a pipe, and we may therefore write, if $\mathfrak{h}$ is the pressure lost, p the absolute pressure of the air, and r its velocity—

$$\mathfrak{h} = \zeta_a p r^2$$

where ζ_a is the coefficient of resistance to be determined by experiment. Some experiments were made by Professor Riedler on these draining tanks. The following table gives the results and the calculated value of ζ_a.

Absolute pressure p, lbs. per sq. in.	Velocity of air in main, ft. per sec.	Observed loss of pressure at one draining tank, $\mathfrak{h}$	Calculated value of ζ_a
104·6	19·5	0·94	0·00002363
104·1	19·0	0·95	0·00002527
84·1	28·5	2·42	0·00003543
106·9	24·25	2·14	0·00003404
102·0	18·25	1·00	0·00002943
		Mean	0·00002956

These figures are fairly accordant, and the loss at each draining tank may be taken as—

$$\mathfrak{h} = 0·00003 \, p r^2 \qquad (11)$$

Experiments by Professor Riedler and Professor Gutermuth on the Resistance of the Paris Air Mains.—The old air mains of Paris consist of cast-iron pipes 30 centimetres = $11\frac{3}{4}$ inches = 0·98 foot in diameter. It was on these that the experiments were made. For the new station wrought-iron pipes of 20 inches diameter are being used. The importance of these experiments lies chiefly in the large scale on which they were carried out. The volumetric efficiency of the compressors was first determined. Then, at different points on the pipe-line, the pressure was observed when the compressors were running at known speeds. On Sundays the whole pipe-line could be coupled up, all work in the city being stopped. The air then passed through the southern main and back by the northern main, a distance of 10 miles. The main is somewhat complicated, having been constructed piecemeal during the gradual extension of the enterprise. In the 10-mile length there are four draining tanks, twenty-three draining traps or siphons, and forty-two stop-valves. The resistance of these is included in the observed results. It appears, however, that the most serious additional resistance was that of the draining tanks, and as special experiments were made on these, their resistance can be estimated with very approximate accuracy and allowed for.

The table opposite gives the results of the experiments, reduced to English measures. The columns marked with an asterisk are those taken directly from Professor Riedler's tables. The others are deduced from his figures.

The last column gives the values of ζ calculated by equation (10).

It will be seen that the values of ζ are fairly consistent considering the difficulties of the observations. The discrepancies are irregular, and obviously due to inconsistencies in the data of the experiments. For instance, experiments XIII and XIV were made on the same main at nearly the same initial pressure. In XIV the velocity was sensibly greater than in XIII, but the loss of pressure is much greater in XIII than in XIV.

Professor Riedler appears to believe that his experiments on the Paris mains show that the friction of air in pipes is considerably less than it was believed to be, from the results of earlier experiments on a smaller scale. The Author does not think that this is so, but that, on the contrary, when properly reduced they are consistent with previous results, and that all the known experiments fairly support each other. In a paper on the 'Coefficient of Friction of Air Flowing in Long Pipes,'[1] it was shown that Mr. Stockalper's experiments gave the following results.

		$\zeta =$
Mean for 0·492-foot pipe .		0·00449
,, 0·656-foot pipe .		0·00377

These experiments, as well as some others given in that paper, show that the value of ζ for air, as is the case for water, decreases as the size of the pipe is larger. Though experiments on the flow of air are not numerous enough to furnish any very trustworthy law, the author gave in 1880 the following expression for the value of ζ :—

$$\zeta = 0\cdot0027 \left(1 + \frac{3}{10d}\right)$$

[1] *Minutes of Proceedings Inst. C. E.*, vol. lxiii. p. 29 .

EXPERIMENTS ON THE RESISTANCE OF THE PARIS AIR MAIN

No.	Position of pipe	Length in miles l	Initial pressure in lbs. per sq. inch p_1 (absolute)	Terminal pressure in lbs. per sq. inch p_2 (absolute)	$p_1 - p_2$	Number of draining tanks	Calculated resistance of the draining tanks in lbs. per sq. inch	Terminal pressure corrected by deducting resistance of tanks	Corrected value of $p_1 - p_2$	Volume of air passing per second reckoned at atmospheric pressure in cub. ft.	Volume of air per second at initial pressure p_1 in cubic feet	Initial velocity in feet per second v_1	ζ
I.	Entire pipe line	10·28	106·6	77·2	29·4	4	8·95	86·15	20·45	135·0	18·62	21·69	0·00229
II.	„	10·28	114·4	78·8	35·6	4	8·45	87·25	27·15	135·5	17·42	23·09	0·00316
III.	„	10·28	119·4	100·6	18·8	4	4·76	105·36	14·04	109·2	13·44	17·32	0·00280
IV.	„	10·28	116·1	107·0	9·1	4	2·44	109·44	6·66	79·1	10·01	13·31	0·00255
V.	„	10·28	114·4	108·8	5·6	4	1·62	110·42	3·98	64·5	8·29	10·99	0·00229
VI.	Rue de Charonne	2·74	116·1	111·7	4·4	1	0·95	112·65	3·45	98·6	12·48	16·55	0·00327
VII.	„	2·74	119·1	116·9	2·5	1	0·42	117·32	2·08	66·4	8·17	10·83	0·00149
VIII.	„	2·74	114·4	105·4	9·0	1	1·83	107·23	7·17	135·5	17·42	23·09	0·00345
IX.	„	2·74	119·4	113·2	6·2	1	1·14	114·34	5·06	109·2	13·44	17·82	0·00397
X.	Fontaine—Station 1	2·08	102·9	100·5	2·4	1	0·95	100·50	2·40	109·2	15·80	20·68	0·00214
XI.	Charonne—Fontaine	5·45	112·5	108·6	3·9	2	1·27	109·87	2·63	79·1	10·34	13·71	0·00184
XII.	Fontaine—Station I	2·08	107·7	107·2	0·5	—	—	107·20	0·50	79·1	10·79	14·31	
XIII.	Station—Rue de Charonne	2·75	113·9	110·3	3·6	1	0·71	111·01	2·89	84·2	10·86	14·40	0·00365
XIV.	„ „	2·75	106·3	105·8	2·5	1	0·84	106·64	1·66	89·3	12·12	16·06	0·00181
								Mean					0·00290

Experiment XII. gives a result so anomalous that it is impossible not to suspect some error of observation or record.

This gives the following values for ζ:—

	$\zeta =$
Pipe 0·492 feet diameter .	0·00435
„ 0·656 „ „	0·00393
„ 0·980 „ „	0·00351

which are only a little different from those deduced from Stockalper's and Riedler's experiments.

Rounding off slightly the mean of Professor Riedler's results, it will be assumed in the remainder of this chapter that, for air flowing in pipes of not less than 1 foot diameter, $\zeta = 0·003$. Putting this value in equations (7a) and (9)—

$$r_1 = \sqrt{\left\{74,300,000 \frac{d}{L} \frac{p_1{}^2 - p_2{}^2}{p_1{}^2}\right\}} \quad . \quad (12)$$

$$p_2 = p_1 \sqrt{\left\{1 - \frac{r_1{}^2 L}{74,300,000d}\right\}} \quad . \quad . \quad (13)$$

Loss of Pressure per Mile of Pipe.—In order to indicate what kind of results this formula leads to, the following cases have been calculated.

1. Given the diameter of the pipe, and the initial pressure and velocity of the air entering the main, it is required to find the loss of pressure in one mile of transmission. By simple transposition, putting $L = 5,280$ feet—

$$p_2 = p_1 \sqrt{\left\{1 - \frac{r_1{}^2}{11,072d}\right\}} \quad . \quad . \quad (14)$$

Assuming initial velocities of 25, 50, and 100 feet per second, and initial pressures of 50, 100, and 200 lbs. absolute, the following values are obtained :—

Diameter of pipe, feet	Initial velocity r_1 ft. per sec.	Values of terminal pressure p_2, when			Percentage of initial pressure lost in one mile
		$p_1 = 50$	$p_1 = 100$	$p_1 = 200$	
1·0	25	48·8	97·7	195·4	2·4
—	50	45·3	90·6	181·2	9·4
—	100	26·9	53·8	107·6	46·2
2·0	25	49·4	98·9	197·8	1·2
—	50	47·7	95·4	190·8	4·6
—	100	40·1	80·3	160·6	19·8

The percentage of pressure lost in a mile is the same whatever the initial pressure. It must not, however, be assumed that the loss in two miles is double the loss in one mile. The velocity increases and the density diminishes going along the main.

It is clear that when the velocity initially in the main exceeds 50 feet per second, the loss of pressure becomes serious even in a distance of a mile. How far that involves a loss of efficiency will be considered later.

2. Another mode of looking at the question is interesting. Suppose the percentage loss of pressure is assumed and the initial velocity calculated. In the following table this has been done for transmissions to distances of 1, 5, 10. and 20 miles.

Length of main in miles	Diameter of main in feet	Initial velocity v_1 when the percentage loss of pressure in transmission is				
		1	2½	5	10	20
1	1	16·7	26·3	37·0	52·5	73·1
	2	23·7	37·3	52·4	71·18	100·6
5	1	7·5	11·8	16·5	23·5	31·8
	2	10·6	16·6	23·3	33·2	44·9
10	1	5·3	8·3	11·3	16·6	22·5
	2	7·4	11·7	15·9	23·4	31·8
20	1	3·7	5·9	8·3	11·8	16·0
	2	5·4	8·3	11·7	16·6	22·6

Transmission is possible to the longest distances here assumed at velocities not impracticably low, with losses of pressure which would not hinder efficient use of the air.

Action of the Air Compressors.—When an air compressor is driven by a steam engine, there is a difference between the work done by the steam and the work done on the air, due to the work expended in friction of the mechanism, and measured by the difference of area of the indicator diagrams of the steam cylinder and compression cylinder. If the compressor is driven by water power there will also be a corresponding loss in friction of the mechanism, probably not widely different in amount. It will be sufficient here to consider a compressor driven by steam.

Let u be the work expended, measured on the steam-cylinder indicator diagram, and u_1 the corresponding work shown on the compressor-cylinder diagram. Then, if η_1 is the efficiency of the mechanism,

$$u_1 = \eta_1 u.$$

It is convenient at present to take u and u_1 to be the work in foot-pounds per pound of air compressed. In Professor

Kennedy's tests of some of the older compressors at Paris, it
appeared that $\eta_1 = 0.845$. Experiments by Professor Gutermuth
on the new Riedler compressor gave $\eta_1 = 0.87$, a result not widely
different.

In all compressors there are some losses of work due to
clearance, to imperfect action of the valves, to leakage, and to
other causes. But unless the machine is badly constructed
these need not be large. If the air is compressed to a pressure
above that in the mains—as happens in some cases—there
is a loss, for the unbalanced expansion uselessly heats the air;
but this again can be kept within narrow limits if the valves act
promptly and properly. The chief loss of work in compressors
is due to useless heating of the air. It is not, of course,
impossible that some of the heat thus generated should be usefully employed. But to make use of it would involve complica-
tions. It is said that in Birmingham, where the distance of
transmission was not great, some of the heat reached the motors,
and then there was an economy in the amount of air used. But
practically, in most cases, heat given to the air in the compressor
is in fact wasted before the air is used for motive purposes. A
perfect compressor for power distribution would be one in which
the air, taken in at atmospheric pressure p_a, should be compressed
isothermally to an absolute pressure p_1 equal to that in the
mains at the compressing station, and then delivered into the
mains without valve-resistance.

Such a machine, working without friction, or clearance, or
valve-losses, would require for each pound of air compressed an
amount of work given by the equation—

$$U_2 = P_a V_a \log_e \frac{p_1}{p_a} \text{ foot-lbs.,}$$

where P_a is the atmospheric pressure in lbs. per square foot, and
V_a the volume of 1 lb. of air at that pressure.

If, as hitherto assumed, the initial temperature of the air is
60° Fahrenheit,

$$U_2 = 27,710 \log_a \frac{p_1}{p_a}.$$

Fig. 62 shows the indicator diagram of such an engine.
O A D E is the work done by the atmosphere on the piston in
the suction stroke. O B C D E is the work done on the air in

the compressing stroke. The work expended in compression is
therefore the shaded area, A B C D.

If the air is compressed at constant temperature, the compression curve is the isothermal D F. The work of compression
is then the area A D F B. If the air is compressed without any
cooling during compression, the compression curve is an
adiabatic such as D G. The work expended in compression is the

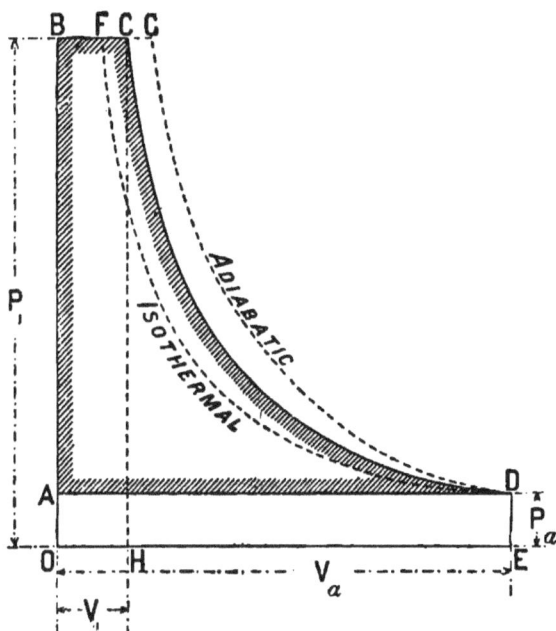

FIG. 62.—ACTION OF COMPRESSOR.

area A D G B. Consequently, in that case the area F G D represents the work expended in useless heating of the air. This
is easily calculated, for it can be shown that the work expended
per pound of air in adiabatic compression is :—

$$\frac{\gamma}{\gamma-1} \, P_a V_a \left[\left(\frac{p_1}{p_a} \right)^{\frac{\gamma-1}{\gamma}} - 1 \right]$$

Putting $\gamma = 1\cdot41$ and $P_a V_a = 27,710$ as before,

$$U'_2 = 95,630 \left[\left(\frac{p_1}{p_a} \right)^{0\cdot29} - 1 \right] \qquad . \qquad . \qquad (15)$$

Q

If $\eta_2 = U_2/U'_2$, then η_2 may be termed the efficiency of the compressing process. For simple adiabatic compression, neglecting clearance and friction losses—

$$\eta_2 = \frac{\log_e \frac{p_1}{p_a}}{3\cdot451 \left[\left(\frac{p_1}{p_a} \right)^{0\cdot29} - 1 \right]}$$

This may be called the theoretical value of η_2.

p_1	p_a	$\frac{p_1}{p_a}$	Theoretical value of η_2
44·1	14·7	3	0·85
73·5	—	5	0·78
102·9	—	7	0·74
147·0	—	10	0·70

It is clear, therefore, that if the air is not cooled during compression, the efficiency of the process decreases as the pressure to which the air is compressed is greater. Now in most cases the cooling arrangements are very imperfect, and the compression is nearly adiabatic. Consequently there has been reluctance to use high initial pressures, and this diminishes the facility for

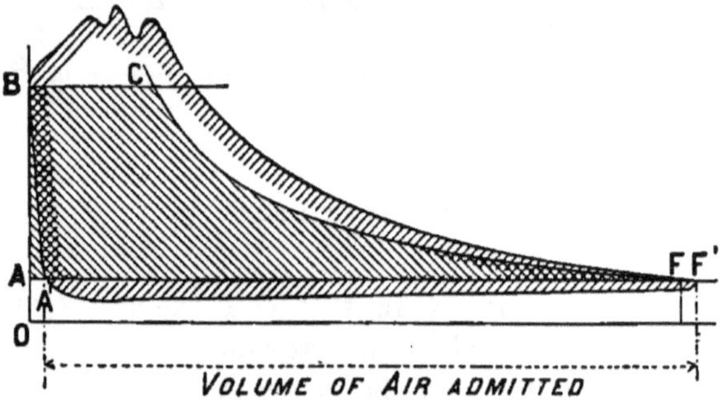

Fig. 63.—Colladon Compressor. Revolutions, 104; Work wasted, 38·15 per cent. of Useful Work.

distributing power by compressed air. As the other losses besides that due to heating are serious, the whole efficiency of the compressor is considerably less than the values calculated above.

Fig. 63 shows an actual diagram from one of the best of the

older Colladon compressors. The shaded area, B C F A, is the useful work of compression, U_2, corresponding to the area A B C D in fig. 62. The thick line is the actual indicator diagram. It will be seen that the actual diagram is 38·15 per cent. larger than the isothermal diagram for the volume of air admitted, the difference being due partly to the waste of work in heating the air, and partly to valve resistance, causing a loss of pressure in the suction stroke and an excess pressure during delivery into the mains. In this case η_2 measured from the diagram is 0·72. In several compressors tried by Professor Riedler the loss of work in the compressing process was twice

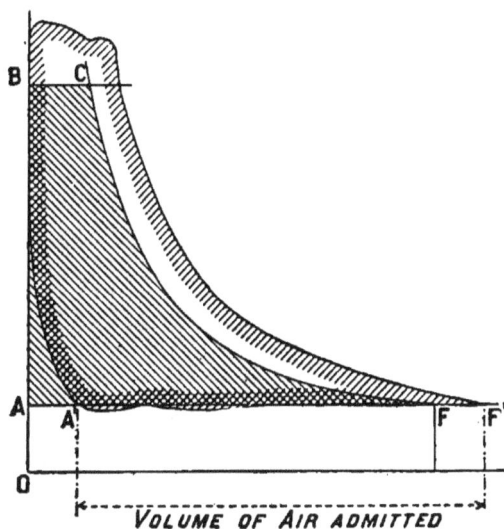

FIG. 64.—COCKERILL COMPRESSOR.

as great as in this case. Fig. 64 shows a similar diagram for one of the Cockerill compressors at Paris. The excess work here is 40·2 per cent. of the useful work, and consequently $\eta_2 = 0.71$.

Obviously, part of this loss can be saved if the air is cooled during compression, and all compressors are provided with some means of cooling. Very generally a water jacket to the compressing cylinder is used, but the action of this is very imperfect. At Birmingham, although the compressors were well water-jacketed, and though the pressure was only 45 lbs. (gauge), the temperature of the air delivered was about 280° F. The area of surface of the cylinder is small compared with the volume of air compressed, and air parts with heat to a metal surface slowly. Spray injection into the cylinder answers much better in keeping down the temperature. But it is believed that the result is partly deceptive, the cooling going on after the air is completely compressed, so that the compression curve is hardly so much flattened as might be expected from the temperature at which the air is delivered.

Fig. 65 shows a diagram from a compressor with water

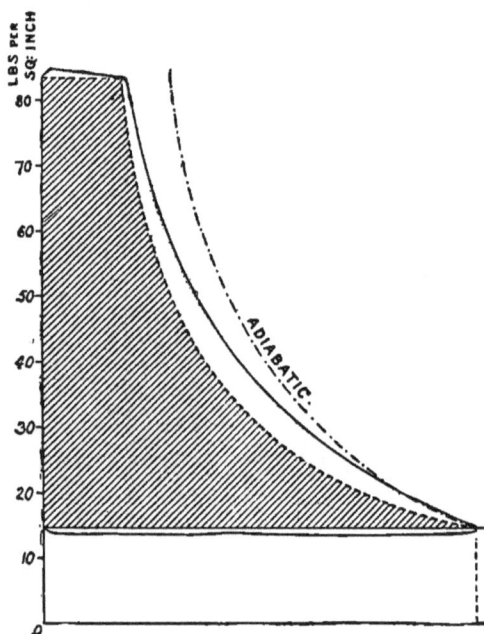

FIG. 65.

pistons. Here the cooling is much more effective. The piston speed is 180 feet per minute. The excess work is only 19·6 per cent. of the useful work, and $\eta_2 = ·84$.

The only remaining means of reducing the loss of work by heating is to compress in two or more stages, and cool the air thoroughly between the stages. This intermediate cooling can be easily effected, the air being taken through tubular vessels presenting any amount of cooling surface that may be required.

In fig. 66 A B K F is the diagram of isothermal compression, as before. F E G is an adiabatic, so that in single-stage com-

pression, without cooling, K D F G represents the work lost in
useless heating. Now let it be supposed that the compression
is effected in two stages. In the first stage, no cooling being
assumed, the work lost will be the area H E F A. But the air
is then cooled to its initial temperature, and the volume shrinks
from H E to H D. D is a point on the isothermal F K. During
the second stage D C is the adiabatic compression curve, and
D K C the work wasted in heating the air. It can now be seen
that the effect of the intermediate cooling is to reduce the work
expended in compression by the area C D E G a very material
quantity.

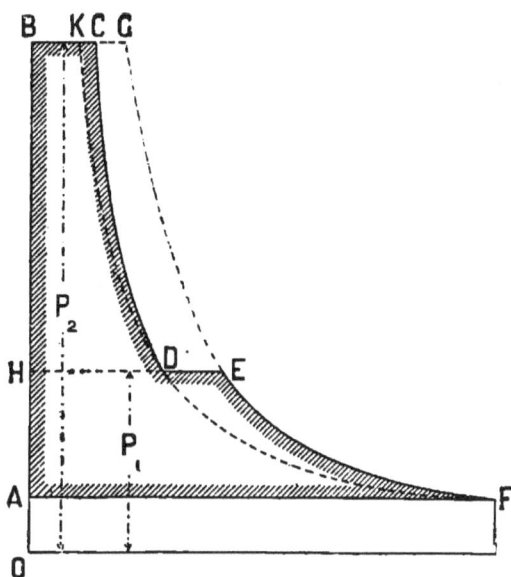

FIG. 66.

It is this stage compression with intermediate cooling, and
with spray injection also, which Professor Riedler has adopted
in the new compressors for Paris, with marked increase of
efficiency. Fig. 67 shows an actual diagram from one of the
Riedler compressors, and it will be seen that the excess compression is only 12·07 per cent., so that in this case η_2 has the
high value 0·89.

Fig. 68 gives the diagram of another Riedler compressor at
Paris, running at a higher speed.

In using air for transmitting power it is important, with a view of diminishing the size of the mains, to adopt high initial

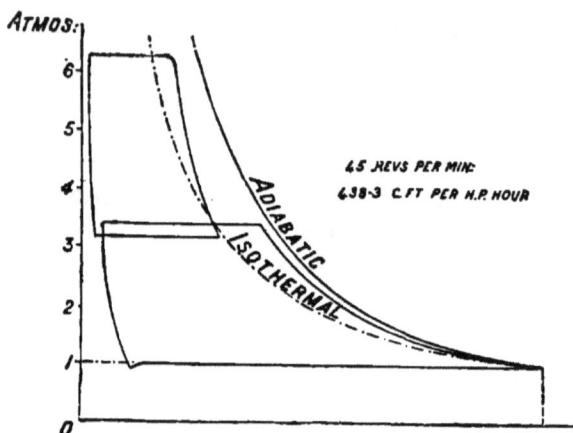

45 REVS PER MIN:
438·3 C. FT PER H.P. HOUR

ADIABATIC

ISOTHERMAL

FIG. 67.

pressures, and it is the inefficiency of the compressing process in ordinary machines which has hitherto prevented the adoption of such pressures. Provided the efficiency of the compressors can be increased, much greater pressure loss can be permitted in

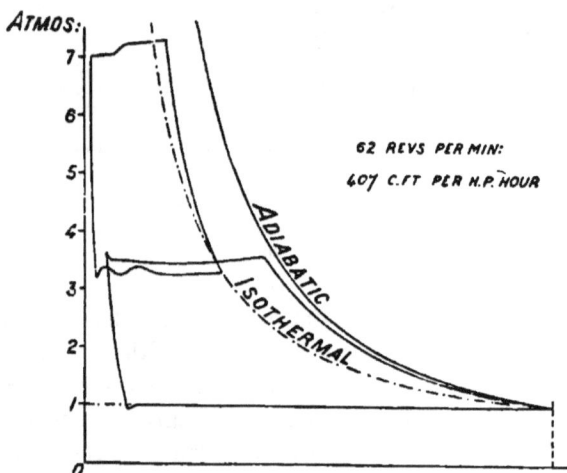

62 REVS PER MIN:
407 C. FT PER H.P. HOUR

ADIABATIC

ISOTHERMAL

FIG. 68.

the mains, and consequently much smaller mains will suffice to transmit a given amount of power.

Professor Riedler and Professor Gutermuth's Experiments on the Efficiency of Compressors.—If, as above, $U_1 = \eta_1 \, U$ is the work expended in compression, after deducting the friction of the mechanism, and

$$U_2 = 27,710 \log_\epsilon \frac{p_1}{p_a}$$

is the useful work done in compression, then

$$\eta_2 = U_2 / U_1 = U_2 / \eta_1 U$$

is a coefficient of efficiency of the compressing process, which includes both the waste of work in useless heating and any losses due to clearance, valve resistance, &c.

Professor Riedler has obtained a series of indicator diagrams from different compressor cylinders. On these he has drawn an ideal isothermal diagram, without clearance or valve losses, and for compression to the pressure in the main. The excess of area of the actual diagram over the ideal diagram is the work wasted in the compressing process, and from this η_2 is easily calculated. The following table gives a series of such results:—

EFFICIENCY η_2 OF COMPRESSION

Type of compressor	Pressure in main, p_1 Atmos.	Lost work in per cent. of useful work	η_2
Colladon, St. Gothard . . .	6	105·0	0·488
„ „ . . .	6	92·0	0·521
Sturgeon	3	94·3	0·515
Colladon	4	38·15	0·722
Slide-Valve Compressor . .	5	49·3	0·670
Paxman	6	42·7	0·701
Cockerill	6	40·2	0·713
Riedler Two-stage . . .	6	12·07	0·892

The table shows how very low is the efficiency of some of the older compressors. Even in the compressors in Paris with single-stage compression $\eta_2 = 0\cdot70$. Hence, if the friction of mechanism is taken account of, and η_1 is put at $0\cdot85$, the resultant efficiency of the compressor mechanism and the compressing process is $0\cdot85 \times 0\cdot70 = 0\cdot595$, or less than six-tenths of the indicated work in the steam cylinder is usefully expended in compression. The Riedler two-stage compressor gives a much better result. Taking $\eta_1 = 0\cdot87$, its efficiency is $0\cdot77$.

Professor Riedler has given some other figures, not based, as the above are, on the measurement of single diagrams, from which the efficiency of the Paris compressors can be calculated in a more trustworthy way. The following data are given as the result of a series of experiments with each compressor :—

	For each indicated steam h.p.	
Type of compressors	Cubic feet of air at atmospheric pressure compressed per hour	Pressure in main, lbs. per sq. inch absolute
Paxman machine (Sturgeon compressors)	264·9	102·9
Cockerill machine (Dubois - François compressors)	300·2	102·9
Riedler machine (Two-stage compressors)	367·4	102·9

From these figures the following are calculated :—

	Steam work per hour in foot lbs.	Weight of air compressed in lbs.	Steam work per lb. of air, =U	Calculated work per lb. for isothermal compression, U_2	Result-ant effi-ciency $\eta_1 \eta_2 = \dfrac{U_2}{U}$	η_1	η_2
Paxman	1,980,000	20·24	97,830	53,920	0·551	0·85	0·648
Cockerill	—	22·94	86,340	—	0·624	0·85	0·735
Riedler	—	28·06	70,550	—	0·764	0·87	0·898

These results agree well with those obtained above in a different way.

Action of the Air Motors.—An air motor is simply a reversed air compressor. Hitherto the conditions of efficiency in air motors have received very little attention. In Paris many of those used are of small size, and in these a good efficiency is not to be expected. The best results have been obtained thus far by adapting old steam engines to work as air motors, and this can be done with very little trouble. It is specially desirable that in an air motor the cylinder clearance should be small or the compression sufficient. Probably the greatest source of avoidable waste in air motors has been leakage at the piston. In a steam engine the condensation on the cylinder wall helps to render the piston tight. In an ordinary air engine, the cylinder surface is more or less dry, and the waste with air from leakage at even small apertures is very great.

A considerable economy can be secured by re-heating the air in a simple form of stove before admitting it to the engine. At first sight this seems a complication likely to involve as much trouble as a steam-boiler. That, however, is not at all the case. The re-heating appliances are simple, there is no risk of dangerous explosion, and the amount of heat which it is desirable to give to the air is insignificant compared with that required in raising steam. Professor Riedler tried an old 80-h.p. steam engine in Paris, which had been adapted to work as an air motor, and which was actually giving 72 indicated h.p., with compressed air at $5\frac{1}{2}$ atmospheres pressure. It was using about 31,000 cubic feet (reckoned at atmospheric pressure), or about 2,376 pounds of air, per hour. This air was heated to a temperature of about 300° Fahrenheit by the expenditure of only 15 pounds of coke per hour. On favourable assumptions a steam engine working to the same power would have required ten times this consumption of fuel at least. Re-heating the air has the practical advantage of raising the temperature of exhaust of the motor, and for the amount of heat supplied the economy realised in the weight of air used is surprising. The reason of this is that the heat supplied to the air is used nearly five times as efficiently as an equal amount of heat employed in generating steam.

In certain cases the air-motor cylinder has been jacketed by hot air. This increases again the amount of work obtained per pound of air used. It can hardly be considered a thermodynamically advantageous process, but it may have advantages practically in raising the exhaust temperature. Lately in some cases, water has been injected in small quantity into the air while passing through the re-heating stove. This passes into the engine as steam. It condenses during the expansion, yielding latent heat to the air, and thus raising the temperature of exhaust. Whether it is advisable from a purely thermodynamic point of view may be doubted, but it seems to have practical advantages possibly in lubricating the cylinder and preventing leakage. When steam is employed in this way the expansion curve rises above the adiabatic and becomes nearly an isothermal. The steam may amount to about 5 per cent. of the weight of air used.

Useful Work done by an Air Motor.—Let the air be delivered from the main to the air motor at the pressure P_2 in lbs. per square foot, or p_2 in lbs. per square inch. (1.) Let it be supposed that the air is used in the motor cold, its temperature being taken at 60° Fahrenheit and its volume in cubic feet per lb. being V_2. Expanded adiabatically in an engine down to atmospheric pressure, p_a, the work done would be in foot lbs. per lb. of air.

$$U_3 = \frac{\gamma}{\gamma-1} P_2 V_2 \left[1 - \left(\frac{p_a}{p_2}\right)^{\frac{\gamma-1}{\gamma}}\right]$$

$$= 95,600 \left[1 - \left(\frac{p_a}{p_2}\right)^{0.29}\right] \qquad . \qquad . \qquad (16)$$

The actual work obtained in any given motor using air cold will be less than this in consequence of incomplete expansion, valve resistance, clearance, leakage, and other losses. Let the actual work shown on the indicator diagram of the air motor be U_4. Then if

$$U_4 = \eta_3 U_3,$$

η_3 is the efficiency with which the fluid is used in the particular air motor, a coefficient which must be determined by experiment.

(2.) Let it be supposed that the air, arriving in the main with the temperature T_2 (absolute), is re-heated to a temperature T_3 before being used in the motor. As it is re-heated at constant pressure, the amount of heat to be given to each lb. of air is 183 $(T_3 - T_2)$ foot lbs., or $0.2375 (T_3 - T_2)$ thermal units. Then the work of adiabatic expansion from a pressure p_2 to atmospheric pressure p_a will be—

$$U'_3 = \frac{\gamma}{\gamma-1} P_2 V_2 \frac{T_3}{T_2} \left[1 - \left(\frac{p_a}{p_2}\right)^{\frac{\gamma-1}{\gamma}}\right]$$

$$= 95,600 \frac{T_3}{T_2} \left[1 - \left(\frac{p_a}{p_2}\right)^{0.29}\right] \qquad . \qquad . \qquad (17)$$

Generally the work shown by the indicator diagram of any engine will be less than this, for the causes mentioned above, and if η_3 is the efficiency with which the motor uses the fluid, the indicated work will be—

$$U'_4 = \eta_3 U'_3$$

There is no reason for expecting η_3 to be different in this case from what it was in the previous one, unless hot-air jacketing, or steam injection, is used. In that case η_3 will be larger, and may be taken to include the additional work due to heat supplied during the stroke.

There is yet one more source of loss in the motor, the friction of the mechanism. If η_4 is the efficiency of the mechanism, then—

$$U_5 = \eta_4 \ U_3 \ \text{or} \ \eta_4 \ U'_4$$

is the effective or brake work of the motor per lb. of air used.

Experiments on Air Motors by Professor Riedler and Professor Gutermuth.—In the following tables some of the experiments are quoted, together with the work per lb. of air, and the values of η_3 and η_4. When Professor Riedler does not give the indicated, but only the brake horse power of the motor, it will be assumed that $\eta_4 = 0.85$ in order to calculate a probable value of η_3.

It will be seen that in the older small motors the efficiency with which the fluid is used ranges from 0.37 to 0.44, which perhaps, considering the kind of motor, is a good result. In the later machines, arranged to work expansively, the efficiency ranges from 0.58 to 0.87, results remarkably good for motors so small. The coefficients are rather higher with re-heated air, showing that the work done is increased even in a rather higher ratio than T_3/T_2.

Some other experiments on small motors may be passed over in order to consider some experiments on an old Farcot steam engine with Corliss valves, which had been converted for use as an air motor.

This engine was nominally of 80 h.p., and worked at 72 i.h.p. in the trials. In all cases the air was re-heated before use to about 300° Fahr. The cylinder was also jacketed by the hot air on its way to the cylinder chest.

The efficiency is therefore 0.81, an extremely good result.

Of all the work obtainable by the expansion of 1 lb. of air received at 95.5 lbs. per square inch, and at a temperature of 300° Fahr., four-fifths is obtained as effective work on the brake. However, part of this work recovered is borrowed from the hot air before admission to the cylinder and given back

ESP

SMALL ROTATING AIR MOTORS

Nominal size	Revolutions per minute	Brake h.p.	Temperature of the air — At admission (Fahrenheit)	At exhaust (Fahrenheit)	Initial pressure, lbs. per square inch absolute	Air used in cubic feet per brake h.p. per hour	Air used in lbs. per brake h.p. per hour	Effective work in foot lbs. of 1 lb. of air = U_3	Calculated work of 1 lb. of air expanded adiabatically = U_2	$\frac{U_3}{U_2} = \eta_3\eta_4$	Assumed η_4	η_3
H.P.						*Old Motors without Expansion*						
½	234	0·506	59	0	73·5	2,330	177·7	11,140	35,670	0·312	0·85	0·368
1	187	1·08	59	5	73·5	1,946	148·5	13,340	35,670	0·374	0·85	0·441
½	241	0·610	122	—	73·5	1,671	127·5	15,530	45,100	0·344	0·85	0·405
1	230	1·42	140	74	73·5	1,625	124·0	15,970	46,500	0·343	0·85	0·404
						Newer Motors with Expansion						
1	221	1·24	60	-33	73·5	1,470	112·2	17,650	35,670	0·495	0·85	0·583
1	207	1·18	59	-44	73·5	1,286	98·2	20,170	35,670	0·565	0·85	0·665
2	200	2·30	68	-51	73·5	1,060	80·9	24,470	36,210	0·676	0·85	0·796
1	210	1·20	158	38	73·5	961	73·3	27,020	42,380	0·638	0·85	0·751
2	235	2·72	158	—	73·5	848	64·7	30,610	42,380	0·722	0·85	0·850
2	241	3·22	140	28	73·5	848	64·7	30,610	41,150	0·744	0·85	0·876

TRIALS WITH AN 80-H.P. FARCOT STEAM ENGINE USED AS AN AIR MOTOR

Initial pressure, lbs. per square inch absolute	Indicated h.p.	Temperature of the air — At admission	At exhaust	Air used in cubic feet per hour — Per indicated h.p.	Per brake h.p.	Lbs. of air per hour — Per indicated h.p.	Per brake h.p.	Indicated work per lb. of air = U_4	Brake work per lb. of air = U_5	Work per lb. of air expanded adiabatically = U_2	$\frac{U_5}{U_4}=\eta_4$	$\frac{U_4}{U_2}=\eta_3$
95·5	72·3	264	70	469	517	35·7	39·5	55,460	50,130	62,120	0·90	0·89
95·5	72·3	305	84	437	475	33·3	36·2	59,460	54,700	65,630	0·92	0·90
95·5	72·3	320	95	425	465	32·4	35·5	61,120	55,770	66,920	0·91	0·91
95·5	65·0	338	120	438	477	33·5	36·4	59,100	54,400	68,470	0·92	0·87

to it by the jacket. That the jacket considerably affected the working during expansion is shown by the temperature during exhaust. The following Table gives the temperatures the air would have reached by adiabatic expansion, and the actual temperatures of the exhaust in the experiments above :—

Initial temperature		Final temperature for adiabatic expansion		Actual final temperature Fahrenheit	Increase of temperature of exhaust due to jacket
Fahrenheit	Absolute	Fahrenheit	Absolute		
264	725	−75	386	+ 70	145
305	766	−53	408	+ 84	137
320	781	−45	416	+ 95	140
338	799	−35	426	+120	155

Practical Calculations on the Distribution of Power by Compressed Air.—It will be convenient first to give a summary of the formulas required in settling a system of compressed air distribution, and afterwards to discuss some special cases.

Let P_a be the pressure in lbs. per sq. ft., p_a the pressure in lbs. per sq. in., v_a the volume of a lb. in c. ft., T_a the absolute temperature of air admitted from the atmosphere to the compressor.

P_1, p_1, v_1, T_1 the same quantities for air discharged from the compressor into the main.

P_2, p_2, v_2, T_2 the same quantities for air arriving at the point of consumption in the main.

U = indicated work done by the steam on the piston of the compressor reckoned in h.p.

$U_1 = \eta_1 U$ = corresponding indicated work on the air in the compressor cylinder.

$U_2 = \eta_2 U_1$ = useful work of compression, or work of isothermal compression from p_a to p_1.

U_3 = available work of air arriving at the motor, or work of adiabatic expansion from p_2 to p_a.

$U_4 = \eta_3 U_3$ = indicated power of the air motor.

$U_5 = \eta_4 U_4$ = effective or brake work of the air motor.

Taking $P_a = 2116\cdot3$, $p_a = 14\cdot7$, $v_a = 13\cdot09$, $T_a = 621°$.

$$P_a v_a = 27,710 \qquad . \qquad . \qquad . \qquad . \qquad . \qquad (1)$$

If w lbs. of air are compressed per second by U h.p. in the steam cylinder corresponding to U_1 i.h.p. in the compressor cylinder, then, allowing both for friction of mechanism, and clearance, leakage, and other losses in compression—

$$550\eta_1\eta_2 u = 27{,}710 \text{ w log}_e \frac{p_1}{p_0}$$

$$w = \frac{\eta_1\eta_2 u}{50\cdot4 \text{ log}_e \frac{p_1}{p_0}} \qquad . \qquad . \qquad . \qquad . \qquad (2)$$

When, by conduction in the main, the air is again at its initial temperature T_a—

$$P_1V_1 = 27{,}710$$

$$p_1v_1 = 192\cdot3 \qquad . \qquad . \qquad . \qquad . \qquad (3)$$

If v_1 is the initial velocity of the air in the main, the diameter of which in feet is d—

$$\frac{\pi}{4}d^2v_1 = wv_1 = 192\cdot3 \text{ w}/p_1$$

$$d = 15\cdot64 \sqrt{\frac{w}{p_1v_1}} \qquad . \qquad . \qquad . \qquad (4)$$

The pressure falls in a main of length l, in feet, from p_1 to p_2, the amount being given by the equation—

$$\frac{p_2}{p_1} = \sqrt{\left\{1 - \frac{v_1^2 l}{74{,}300{,}000 d}\right\}} \qquad . \qquad . \qquad (5)$$

The available work of the air arriving at the motor with the pressure p_2 at atmospheric temperature T_a is, if it is used without re-heating, in h.p.—

$$u_3 = \frac{95{,}600 \text{ w}}{550} \left[1 - \left(\frac{p_a}{p_2}\right)^{0\cdot29}\right]$$

$$= 173\cdot5 \text{ w} \left[1 - \left(\frac{p_a}{p_2}\right)^{0\cdot29}\right] \qquad . \qquad . \qquad (6)$$

If re-heated to T_3 before admission to the motor—

$$u'_3 = 173\cdot5 \text{ w} \frac{T_3}{T_a} \left[1 - \left(\frac{p_a}{p_2}\right)^{0\cdot29}\right] \qquad . \qquad . \qquad (7)$$

Indicated work of motor $= u_4 = \eta_3 u_3$ . . . (8)

Brake or effective work of motor $= u_5 = \eta_4 u_4$. (9)

Values assumed for the Coefficients of Efficiency.—It appears that the efficiency of the mechanism of compressors is from 0·85 to 0·87. In the following calculations it will be assumed that $\eta_1 = 0\cdot85$. As to the efficiency of the process of compression, this varies greatly with the type of compressor. In some of the older single-stage compressors it is as low as 0·5. But taking the best of those tried by Professor Riedler and slightly rounding off the values—

For single-stage compressors .   .   . $\eta_2 = 0\cdot7$

„   two-stage   „   .   .   . $\eta_2 = 0\cdot9$

The loss in the main must be calculated for each special case.

For the air motors, as it is not intended to discuss the use of air on a small scale, it will be assumed that $\eta_3 = 0\cdot85$ and $\eta_4 = 0\cdot9$. These values are slightly below those obtained in the experiments, and the results should therefore be such as are practically realisable in ordinary work.

CASE I.—10,000 h.p. is to be transmitted a distance of 2 miles. Taking $\eta_1 = 0\cdot85$ and $\eta_2 = 0\cdot9$ (two-stage compression) the useful work of compression will be 7,650 h.p. Now, let the air be compressed to 4, 8, and 12 atmospheres on the gauge, then the weight of air compressed will be as follows :—

Case I	Initial gauge pressure, atmospheres	Initial pressure, lbs. per square inch absolute	$\dfrac{p_1}{p_a}$	Weight of air compressed in lbs. per second $= W$
a .	4	73·5	5	94·54
b .	8	132·3	9	69·10
c .	12	191·1	13	59·17

To ascertain suitable diameters for the main, let initial velocities of 30, 50, and 75 feet per second be assumed. For these velocities, and in the cases described, the diameters required will be as follows :—

Case I	Initial pressure, lbs. per sq. in. absolute	Initial velocity in the main, ft. per second	Diameter of main	
			In feet	In inches
a .	73·5	30	3·23	39
		50	2·50	30
		75	2·05	25
b .	132·3	30	2·06	25
		50	1·60	19
		75	1·31	16
c .	191·1	30	1·59	19
		50	1·23	15
		75	1·01	12

None of these mains are impracticable in size, and for the higher pressures and velocities they are surprisingly small, considering that they are shown to be capable of transmitting 10,000 indicated steam h.p. a distance of 2 miles. It remains to examine whether the loss of pressure is serious.

Case I	Initial pressure, lbs. per sq. in. absolute p_1	Initial velocity in the main, ft. per sec.	Diameter of the main in feet	Terminal pressure in lbs. per sq. in. p_2	Loss of pressure in per cent. of initial pressure
a	73·5	30	3·23	72·0	2·0
		50	2·50	68·0	7·4
		75	2·05	57·3	22·0
b	132·3	30	2·06	128·3	3·0
		50	1·60	116·5	11·9
		75	1·31	82·3	37·8
c	191·1	30	1·59	183·4	4·0
		50	1·23	161·1	15·7
		75	1·01	86·4	54·8

None of these losses would render the utilisation of the power impossible, and it is only at the highest pressures and velocities that the fall of pressure is serious.

Lastly, the power developed at the air motor due to 10,000 indicated steam h.p. in the compressor has to be calculated. The air being delivered at a distance of 2 miles, and used cold, the h.p. obtained will be as follows:—

Case I	Initial pressure, lbs. per sq. in. absolute	Terminal pressure in the main, lbs. per sq. in. absolute	Initial velocity in the main, feet per second	Indicated h.p. of air motor	Brake or effective h.p. of air motor
a	73·5	72·0	30	5,146	4,631
		68·0	50	4,998	4,498
		57·3	75	4,546	4,091
b	132·3	128·3	30	4,755	4,279
		116·5	50	4,600	4,140
		82·3	75	4,007	3,606
c	191·1	183·4	30	4,529	4,076
		161·1	50	4,369	3,932
		86·4	75	3,504	3,154

It is very striking how little the efficiency is affected by considerable changes in the initial pressure and velocity in the mains ; with the exception of two cases, for 10,000 steam indicated h.p. expended at the compressor there is obtained from 4,400 to 5,100 indicated h.p. at the air motor. That is to say,

the efficiency of the whole arrangement, compressor, main and
air motor, when the air is used cold, ranges from 44 to 51 per
cent., and that with mains of quite moderate size.

If the method of re-heating at the motor is resorted to,
which in no case involves any great additional trouble, and which
from the small amount of fuel required can often be carried out
with very little additional expense, the h.p. obtained will be as
follows. Let it be supposed that the air arriving at the motor at
60° Fahrenheit is re-heated to 300° Fahrenheit, as in the case of
the Farcot steam engine, details of which are given above.
Then—

$$\frac{T_3}{T_a} = \frac{761}{521} = 1\cdot46.$$

	Case I	Initial pressure, lbs. per sq. inch	Terminal pressure, lbs. per sq. inch	Initial velocity in the main, feet per second	Indicated h.p. of motor	Brake h.p. of motor
a		73·5	72·0	30	7,511	6,761
			68·0	50	7,297	6,567
			57·3	75	6,638	5,973
b		132·3	128·3	30	6,942	6,248
			116·5	50	6,716	6,045
			82·3	75	5,850	5,265
c		191·1	183·4	30	6,612	5,951
			161·1	50	6,380	5,741
			86·4	75	5,116	4,605

Here again, excepting two cases, the efficiency reckoned on
the indicated power is from 64 to 75 per cent., neglecting the
cost of the fuel for re-heating.

CASE II.—A long-distance transmission may now be discussed. Suppose, as before, that 10,000 i.h.p. is developed
in the steam cylinders of the compressors, and is to be transmitted a distance of 20 miles. Taking the same pressures
as before, the calculations of the weight of air compressed
need not be repeated. The initial velocities previously assumed
will, however, be excessive, because the velocity in the main
increases as the pressure falls, and in this case, as the fall will
be considerable, the terminal velocities are much greater than
the initial velocities. Initial velocities of 20, 35, and 50 feet
per second will be assumed.

R

Case II	Initial pressure. lbs. per square inch absolute	Initial velocity. feet per second	Diameter of main in feet	Diameter of main in inches
d . . .	73·5	20	3·96	47·5
		35	2·99	35·9
		50	2·51	30·1
e . .	132·3	20	2·53	30·4
		35	1·91	22·9
		50	1·60	19·2
f .	191·1	20	1·95	23·4
		35	1·47	17·6
		50	1·23	14·8

None of these mains are impracticable in size. The follow-ing table gives the calculation of the pressure loss :—

Case II	Initial pressure	Initial velocity	Diameter of main in feet	Terminal pressure, lbs. per sq. inch	Loss of pressure in per cent. of initial pressure
d .	73·5	20	3·96	68·0	7·4
		35	2·99	47·6	35·3
		50	2·51	Impossible	—
e .	132·3	20	2·53	116·5	11·9
		35	1·91	12·6	90·4
		50	1·60	Impossible	—
f . .	191·1	20	1·95	160·8	15·8
		35	1·47	Impossible	—
		50	1·23	Impossible	—

It turns out on calculation that some of the cases assumed are impossible. That is, the whole initial pressure is insufficient to give the assumed initial velocity. In one other case, the 1·91 foot main, with a velocity of 35 feet, the loss of pressure in the main is impracticably large. There remain, however, four cases in which neither the size of main nor the loss of pressure in a transmission to a distance of 20 miles is such as to render the transmission impracticable. In three cases the loss is remark-ably small, and the terminal pressure quite suitable for applica-tion. For instance, it appears that the air compressed by 10,000 h.p. to 132·3 lbs. per square inch can be transmitted to a distance of 20 miles in a 30-inch main with a loss of pressure of only 12 per cent.

The power delivered at a distance of 20 miles by air motors

using the compressed air can now be calculated. If the air is used cold, the power obtained will be as follows :—

Case II	Initial pressure, lbs. per sq. in. absolute	Terminal pressure, lbs. per sq. in. absolute	Initial velocity in the main, ft. per second	Indicated h.p. of motor	Brake or effective h.p. of motor
d	73·5	{ 68·0	20	4,989	4,490
		{ 47·6	35	4,017	3,615
e	132·3	116·5	20	4,598	4,138
f	191·1	160·8	20	4,365	3,928

If the air is re-heated to 300° Fahrenheit, as assumed in the previous case—

Case II	Initial pressure, lbs. per sq. inch	Terminal pressure, lbs. per sq. inch	Initial velocity in main, ft. per second	Indicated h.p. of motor	Brake or effective h.p. of motor
d	73·5	{ 68·0	30	7,284	6,554
		{ 47·6	35	5,865	5,278
e	132·3	116·5	20	6,713	6,041
f	191·1	160·8	20	6,373	5,735

Here the efficiency of the whole arrangement—calculated on the indicated power, the air being delivered at a distance of 20 miles, and including all losses—is 40 to 50 per cent. if the air is used cold, and 59 to 73 per cent. if the air is re-heated. The results are based absolutely on efficiencies already obtained in similar cases, and the sole loss neglected is possible leakage in the mains.

Fig. 69 is drawn to scale for Case II e. It is a diagram showing the relation of the work expended and useful work recovered when 10,000 h.p. is transmitted 20 miles in a main 30 inches in diameter. E B G R T D is the work expended in a two-stage compressor, compressing 1 lb. of air from 14·7 to 132·3 lbs. per square inch. By cooling in the mains the volume of the air shrinks from B G to B C. The frictional resistance reduces the pressure to 116·5 lbs. per square inch, when the air has the volume M K determined by the isothermal. The work done by an air motor using the air cold is M K A E. If the air is re-heated to 300° the volume expands from M K to

M L (1.46 times). Then the work of the air motor, using re-heated air, is M L H E, and K L H A is the important gain of work due to re-heating. Of course the areas given by the diagram have to be multiplied by the efficiencies given above.

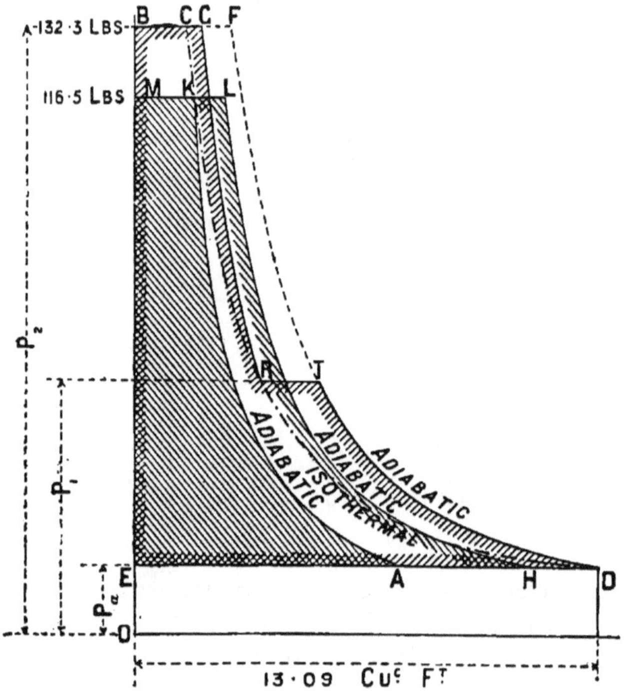

Fig. 69.—Combined Action of Compressor Main and Motor.

In an air main the air expands as the pressure falls. Hence in a very long main the diameter should increase as the pressure falls. An expanding main of this kind has not actually been used.

245

CHAPTER XI

DISTRIBUTION OF POWER BY STEAM

GENERALLY, boilers producing steam are adjacent to the engines generating power. In special cases it has been necessary to convey the steam not inconsiderable distances before it was used in engines. For instance, underground pumping machinery in mines has been worked from steam boilers at the surface by steam conveyed in pipes, protected as far as possible from heat losses. Nevertheless, it would hardly have been thought reasonable to distribute steam widely to considerable distances for power purposes, but for a secondary object. Steam distributed from a central station through pipes can be very conveniently used for heating purposes, as well as for power purposes. The defects of steam, as a means of distributing power, may be balanced by its advantages as a means of distributing heat. At any rate, in the United States the experiment of distributing heat and power from a central station, by steam, has been tried on a very large scale, and with a considerable amount of success.

In 1877, Birdsill Holly patented a system of steam distribution for heating purposes only. The steam was conveyed in pipes, having anchored stuffing boxes at distances of about 100 feet, so that the expansion and contraction of the pipes was provided for. Radiation was diminished by covering the pipes with asbestos and wood. The steam was delivered into the houses through a reducing valve, and used in ordinary heating coils. The condensed steam was discharged through steam traps into the sewers. Generally, the condensed steam, before being discharged, was taken through coils in a chamber, through which fresh air entered the building. The air was warmed, and the condensed steam reduced in temperature to about 100° Fahr.

The heat was thus economised, and the condensed steam discharged at a temperature at which it caused no inconvenience.
Generating and distributing plants on the Holly system were
erected in many towns. In Lockport, for instance, in 1879, the
main steam pipes extended a distance of 16,000 feet.

As early as 1869, Dr. C. E. Emery investigated the problem
of distributing steam in New York.[1] His studies led to the
creation of the largest system of steam distribution hitherto
carried out. Dr. Emery concluded that steam could be economically distributed, from one station, to buildings within a
radius of half a mile. Ten plots of land for stations in New
York were secured, and the construction of the works commenced in 1881. Up to the present two steam stations have
been put in operation, a down-town station, termed Station B,
with boilers working to 16,000 h.p., and an up-town station in
58th Street, designed for boilers of 3,000 h.p., and having about
half this power at work.

The pipes from the down-town station extend through
$5\frac{1}{2}$ miles of streets; those from the up-town station through
$2\frac{1}{4}$ miles of streets.

Down-town Station.—This was erected on an irregular plot
about 75 feet by 120 feet. To obtain space for the boilers, they
are arranged in four tiers, each tier in a separate storey 20 feet
in height. There is a fifth storey for coal storage, and a basement for miscellaneous purposes. Each floor is arranged for sixteen boilers of 250 h.p. each, placed in two rows, facing a central
charging floor. The chimneys are near the centre of the space.
The coal is dumped into small cars in the basement and then
lifted to the top storey, where it is discharged into coal bins.
The coal descends by gravity in chutes to the several floors.
The ashes are discharged down chutes to the basement. The
boilers are Babcock and Wilcox boilers. Fig. 70 shows generally
the arrangement of the building, with one boiler indicated in
place. At this station the pressure is generally maintained at

---

[1] 'District Distribution of Steam in the United States,' *Proc. Inst. C. E.*,
vol. xcvii.; 'The Station B Chimney of the New York Steam Company,'
Trans. Am. Soc. C. E., 1885; 'District Steam Systems,' *Trans. Am. Soc. C. E.*,
1891; 'The Comparative Value of Steam and Hot Water for Transmitting
Heat and Power,' *Trans. Am. Soc. Mech. E.*, vol. viii. All these papers are by
C. E. Emery, Ph.D., of New York, and they contain nearly all the information
available on the subject of steam distribution.

80 lbs. per square inch. Down to 1887, 48 boilers were installed
in three storeys, the fourth
storey being used for coal
storage. The fifth storey
had not been built. In
1891 it appears that four
more boilers had been
placed in the fourth
storey.

Up-town Station.—
This is designed to con-
tain twelve boilers of 200
to 250 h.p. each, of which
half were in place in
1891. The boilers are
on one floor, with coal
storage above and base-
ment below. The plant
was designed to carry
steam of 80 lbs. pressure,
but the pressure has
generally been lower, and
the steam is supplied in
winter only.

The steam distributed
is used chiefly for heating
purposes, but a consider-
able quantity is also supplied from Station B for
power purposes. It is
used for driving steam
engines working printing
presses, electric lighting
machinery, lifts and ven-
tilating machines. In
1887, about 500 steam
engines were supplied
from Station B.

Cost of Steam.—The
charge for steam is based

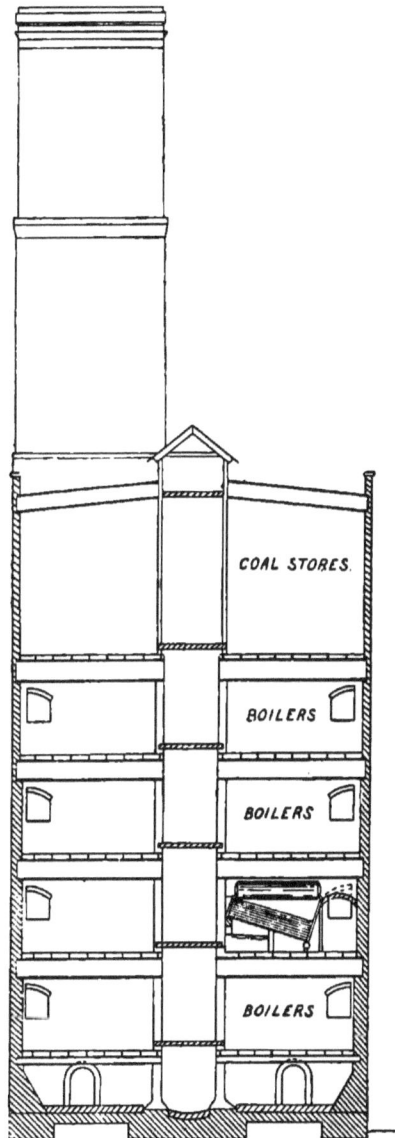

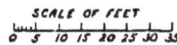

COAL STORES.

BOILERS

BOILERS

BOILERS

SCALE OF FEET
0 5 10 15 20 25 30 35

Fig. 70.

on the quantity of heat supplied. The unit of heat is a 'kal,' which is defined as the heat required to evaporate 1 lb. of water from 100° at 316°, or at 70 lbs. pressure per square inch. One kal is therefore about 1,110 thermal units. Mr. Emery reckons that, on the average, 30 kals per hour give one i.h.p., when the steam is used in ordinary small engines.

When the steam is used for heating only, a rent for the steam supply is paid on an estimate of the heat required, based on a survey of the building. This method has not proved altogether satisfactory. The charge for steam for power purposes, based on the nominal power of the engine, has also proved unsatisfactory. A meter is now used in many cases, which gives a graphic record [1] of the amount of steam delivered, and although not entirely satisfactory, this appears to have been fairly successful.

The ruling price for steam at first was 50 cents per 1,000 kals, which, at 30 kals per i.h.p. per hour, is equivalent to $\frac{3}{4}d.$ per i.h.p. hour. It is stated that now a sliding scale is adopted, the charge being 70 cents per 1,000 kals to small users, and 40 cents to large users. These charges are equivalent to 12l. 10s. and 7l. 10s. per h.p. per year of 3,000 hours. To newspapers and other establishments working chiefly at night the charge is 30 cents per 1,000 kals, equivalent to 5l. 12s. per year per h.p.

Working of the System.—The steam mains are stated to be in as good condition as when laid, though they have been under a continuous steam pressure of 80 lbs. per square inch night and day. On the other hand, the condensed water return mains have proved a failure from external corrosion.

To a great extent they are now disused. Mr. Emery appears to think that the temperature of the steam mains kept them externally dry, but that the return mains, at a temperature of 200° to 212°, did not keep dry, and the corrosion was specially active at that temperature. (Emery, 'District Steam Systems,' p. 193.) With the earlier Holly system, worked at a lower pressure and in smaller towns, it was possible to cool down the exhaust steam to a temperature at which it could be discharged direct into the sewers without mischief. In the larger system

A curve is drawn on a paper strip, the abscissæ representing time and the ordinates the weight of steam flowing.

in New York, where in some cases the steam is supplied to consumers having no control of the basements of the buildings, and
where the steam pipes themselves must be drained at every dip,
it is more difficult to ensure the discharge of the water at a low
temperature.   If the water is discharged at a temperature above
212°, it generates steam when the pressure is released.   Mr.
Emery states that since the use of return mains was discontinued
much steam is finding its way into the sewers, and escaping at
times from man-holes.   Mr. Emery's present opinion is, that
rolled brass return mains should be used, with cast-iron fittings.
If the feed is returned to the boilers at 200° or 212°, he estimates
that there is a saving of 10 per cent. of fuel.   This would make
it pay to use brass return mains.   Mr. Emery believes that the
steam system has been a great public convenience, and that,
where property owners would associate themselves to erect a
steam plant for the improvement of their property, the enterprise would be a remunerative one.   He also thinks that steam
systems for winter use only, the plant being shut down in
summer, would pay well, partly because very cheap coal can be
used.   The New York plant, in spite of some faults almost
inevitable in a new enterprise, is running, and yielding a large
income. Many buildings would have their rental value materially
lowered if the steam supply were cut off.

   The Pipe Mains.—Cast-iron pipes with flanged joints have
been used, but in the United States the pipes are welded
wrought-iron pipes.   In the original Holly system the joints
were stuffing-box joints, to permit free expansion and contraction of the pipes.   The stuffing boxes give great trouble,
even with low pressures, and involve the construction of a great
number of man-holes for access to the joints.   In the New
York supply thick cast-iron flanges are fixed on the ends of
the wrought-iron pipes.   At first the wrought-iron pipe was
fixed in a slightly conical hole in the cast-iron flange by rolling;
the end of the pipe was turned square and abutted against
a turned shoulder.   A caulking space a quarter inch wide
and one inch deep was left at the back of the flange.   The
later pipes used have screwed ends.   A cast-iron flange is
screwed on, and iron cement caulked into a space at the back
of the flange.   Some of the joints have flat faces on the flanges;
others have a spherical joint on the flanges, permitting a

limited variation of direction of the pipes. The joints are made tight by a copper ring smeared with thin red-lead.

At New York the large pipes are 16, 15, 13, and 11 inches external diameter and about a quarter inch thick. The smaller pipes are the ordinary standard wrought-iron piping. For the larger pipes, if p = steam pressure in lbs. per sq. in., d = diameter in inches, t = thickness in inches, then

$$t = c\,p\,d$$

where c is 0·00022 to 0·00026.

Variators or Expansion Joints.—The variation of temperature in steam pipes is very great, and provision must be made

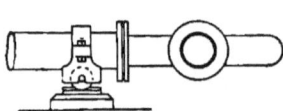
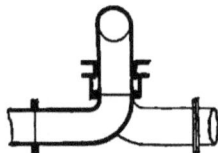
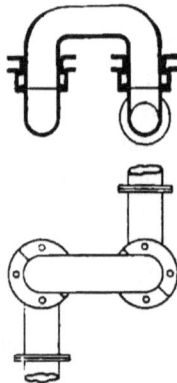
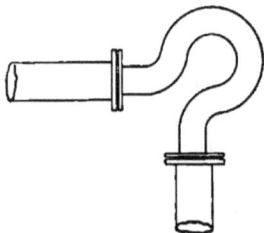

FIG. 71.   FIG. 72.

for permitting expansion and contraction without straining the joints. If the range of temperature amounts to 130° or 150° the pipes will expand and contract $\frac{1}{600}$th to $\frac{1}{750}$th of their length, or, say, 1 inch in 50 to 62 feet of length. Holly used stuffing boxes, but these give great trouble from leakage and the wearing of the packing by the sliding of the pipes. Bends (fig. 71) may be used, but these allow only limited motion. A better plan is to combine bends and stuffing boxes (fig. 72). Then the pipe rotates in each stuffing box without sliding longitudinally, and the wear of the packing is diminished. Flexible diaphragm joints have been tried, but as usually made they permit only a very small amount of motion. If the diaphragm is strong enough to resist the pressure, it is so stiff that

only a small amount of bending is possible without overstraining
the diaphragm.

In the New York supply Dr. Emery has adopted very thin
diaphragm plates of copper, supported by radiating hinged bars.
Thus a flexible and sensitive diaphragm is obtained, perfectly
steam-tight, and the steam pressure is carried by the movable
radiating bars.

Fig. 73 shows one of these variators. The corrugated
copper diaphragm is 0·04 inch thick. The inner and outer

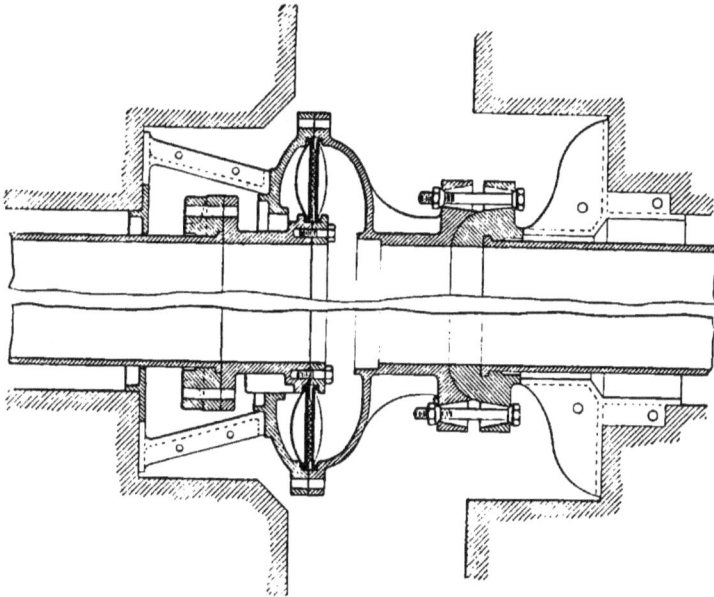

FIG. 73.

edges of the diaphragm are flat, and are clamped between strong
cast-iron plates. On either side of the diaphragm are the loose
bars resting at each end on rough knife edges. Each radiating
plate supports a part of the diaphragm, without in any way
hindering its motion. The length of the radiating plates is
6 inches. The set on one side of the diaphragm resist the steam
pressure; the set on the other side support it, if a vacuum is
accidentally formed in the pipes. The movement at each
variator may amount to $1\frac{1}{4}$ or $1\frac{1}{2}$ inch without overstraining
the diaphragm. A variator is provided for each 50 feet of pipe.

Sometimes for convenience two diaphragms are arranged at one variator. At the elbows and at the variators the pipes are anchored to the brickwork. The service pipes are taken off from the anchored ends of the pipes.

Protection against Radiation Loss.—To prevent radiation from steam pipes various materials are used—straw laid parallel to the pipes and covered with loam, felt, cork, fossil meal, *papier mâché*, slag wool, and asbestos. Slag wool is very effective, and is obtained by blowing steam through molten slag. There seems to be some doubt, however, whether, if the outside of the pipe is moist, it does not increase corrosion. In any case it is desirable to keep the outside of the pipes dry, and this is most likely to be secured if they are kept continuously heated.

Let k lbs. of steam be condensed by radiation per sq. ft. of the surface of the pipe per hour. Then if G_1 is the weight of steam entering and G_2 the weight of steam delivered at the other end of a main of diameter d and length l

$$G_1 = G_2 + \pi d l k.$$

The fraction of the steam condensed and for useful purposes wasted by radiation is

$$\eta = \frac{G_1 - G_2}{G_1} = \frac{\pi d l k}{G_2 + \pi d l k}$$

Obviously the proportion of the steam wasted increases as G_2, the steam used, diminishes. It will be greater as the demand for steam is smaller.

In New York the pipes are laid in brick trenches, the brickwork being kept 4 inches away from the pipes. The pipes are asphalted and the space between pipe and brickwork packed with slag wool. Over the pipes is a roof of short tarred planks bedded in cement. These are covered by tarred paper carried well down the side walls to exclude percolating water. There are usually two intermediate piers with saddles supporting the pipe, in each 50-foot length between the variators. At all dips traps are connected for discharging condensed steam.

Leakage.—It is clear that very considerable difficulties arose in New York from leakage of steam, and repair was difficult, because the steam supply could not be stopped. Dr. Emery attributes these difficulties entirely to bad workmanship.

In 1886, the losses due to leakage were investigated, the return mains being then in operation. With steam for 6,000 h.p. supplied in winter and for 3,000 h.p. in summer it was estimated that the radiation loss amounted to 150 h.p. and the leakage loss to 500 h.p. That is $2\frac{1}{2}$ per cent. and $8\frac{1}{4}$ per cent. on the winter supply and 5 per cent. and $16\frac{1}{2}$ per cent. on the summer supply respectively. In 1887, with nearly double as much steam supplied, the radiation loss was estimated at 350 h.p. and the leakage loss at 720 h.p.

Size of Mains.—The pipes are designed for a velocity of 80 feet per second, and it is believed that at this velocity the pressure loss does not exceed 10 lbs. per sq. in. per half mile of transmission.

Let w = weight of steam flowing per hour in lbs.

d = diameter of main in inches.

Then, Dr. Emery's rule can be put in the form,—

$$w = 100\, d^{\frac{5}{2}}$$

Quantity of Power transmitted by Steam Mains.—Let u be the velocity of the steam in the mains, and v the volume of a pound of steam in cubic feet. Then

$$w = \frac{\pi}{4}\frac{u}{v}\, d^2$$

If $u = 80$ feet per second, which is not excessive considering the small density of steam,

$$w = c\,\frac{\pi}{4}\, d^2$$

where c is a constant depending on the steam pressure.

Gauge pressure, lbs. per square inch	Absolute pressure, lbs. per square inch	v Cubic feet per lb.	c
45	60	7·04	11·36
75	90	4·81	16·63
125	140	3·18	25·15

If 30 lbs. of steam per hour will develop in ordinary engines one i.h.p., then the h.p. transmitted by a main is

$$\text{H.P.} = 120\,w = 94\,c\,d^2$$

Diameter of main in inches	Steam gauge pressure, lbs. per square inch	Weight of steam conveyed, lbs. per second	H.P. transmitted by main
6	45	2·23	267
6	75	3·26	391
6	125	4·93	592
12	45	8·92	1,070
12	75	13·05	1,570
12	125	19·74	2,370
18	45	20·07	2,410
18	75	29·38	3,530
18	125	44·45	5,330

These figures are necessarily rough values. But they show that, with steam, very large amounts of power can be transmitted, without serious pressure loss, through mains of moderate size.

CHAPTER XII

DISTRIBUTION OF GAS FOR POWER PURPOSES

THE distribution of gas in town districts to many consumers for use in generating power involves nothing new or untried. The convenience and cheapness of this method of distributing a means of obtaining power are so remarkable, that a considerable development of the use of gas for power purposes is likely to be effected. Power can be obtained from a gas distribution in large or small quantities, with freedom from many of the drawbacks attending the use of steam power, and at a cost proportional to the amount of power actually used. Two independent systems of supplying gas for power purposes have to be considered. In one ordinary lighting gas is used, and the demand for gas for generating power is in that case an important secondary source of revenue for existing gas companies. In the other a gas of a cheaper description is manufactured and distributed specially for generating power.

In the United Kingdom there has been expended on capital account in gas undertakings, chiefly for lighting purposes, a sum of 60,000,000l. The amount of coal carbonised annually is 10,000,000 tons, and the quantity of gas manufactured is 98,000,000,000 cubic feet. If all this were used for producing power it would furnish about 1,000,000 h.p. for 3,000 working hours in the year. Lighting gas is already in many towns used on a considerable scale for heating and power purposes. Mr. Trewby, the President of the Institute of Gas Engineers, estimates that in the London district alone there are 70,000 gas cooking and heating stoves, and 2,500 gas engines.

Gas has one advantage over electricity or compressed air, namely, that storage can be so cheaply provided that the manufacture can be carried on continuously and uniformly throughout the twenty-four hours.

Use of Lighting Gas for Generating Power.—The cost of ordinary lighting gas is not so great as to preclude its use on a large scale for power purposes. Taking the cost of lighting gas at 2*s*. to 3*s*. per 1,000 cubic feet, and the consumption of gas in an engine with an ordinary varying load at 26½ cubic feet per effective h.p. hour, the cost of the power for gas only is 8*l*. to 12*l*. per effective h.p. per year of 3,000 working hours. For interest on the cost of gas engine and engine house, depreciation and wages of engineer, an allowance of 3*l*. per annum per effective h.p. is sufficient. Hence the total cost of power derived from lighting gas would be from 11*l*. to 15*l*. per h.p. per year of 3,000 hours.

It must be pointed out, however, that the price charged for lighting gas includes interest on a large network of mains, and the loss due to leakage over an extensive area. This part of the cost is not fairly chargeable against gas used for power purposes. In a distribution of gas for power only, the system of mains would be comparatively simple, and it would not be necessary to provide for the wide fluctuations of demand which occur in a distribution for lighting. There would be comparatively few consumers each taking comparatively large quantities of gas. It appears that the cost of manufacturing gas, including coal, wages, and petty stores, is about 10*d*. per 1,000 cubic feet. Probably 18*d*. per 1,000 cubic feet would allow margin enough for profit and cost of distribution, to power users in a manufacturing quarter not unfavourably distant from a generating station. But, at that price, the cost of power for gas only would be 6*l*. per effective h.p. year, and the total cost 9*l*. per effective h.p. year, including interest and depreciation on the cost of gas engine and labour.

The Dessau Central Station for Electric Lighting.—This station was put in operation in 1886, having been erected by the Gas Company with the object of increasing the consumption of gas during the daytime, and at the same time of meeting the demand for electric light. It has since been extended, and it has been successful enough, both mechanically and financially, to show that the production of power from lighting gas on a fairly large scale is practically and commercially possible. The motor installation consisted at first of two 2-cylinder Otto system gas engines, of 60 h.p. each, one single-cylinder engine of 30 h.p., and one of 8 h.p. The group of engines worked up to about 160 effective h.p. The dynamos were driven by belting and counter-shafts. In 1891, an engine of 120 effective h.p. was erected with directly coupled dynamo, and one of the 60 h.p. engines and the 30 and 8 h.p. engines were removed. The jacket water is cooled by air coolers and used over again. The air coolers have 1,076 sq. ft. of cooling surface, an injector worked by pressure water from the town mains being used to circulate air through the cooler. The consumption of water is 5 gallons per h.p. hour for all purposes.

In an electric station, gas engines have the advantage that they can be started and stopped when required, and they have no stand-by losses like those of steam boilers. On the other hand they do not work efficiently except at full load. At Dessau, a large accumulator battery is used for storing energy when the demand would not keep the engines fully loaded, and re-storing it in hours of small demand when the engines are stopped. The efficiency of the battery is 79 per cent. on the average of the year. About 52 per cent. of the whole supply passes through the battery, so that the waste of current in the battery is about 11 per cent. of the total yearly supply.

The average gas consumption (before 1891) was $26\frac{1}{2}$ c. ft. per effective h.p. hour.[1] Motors of varying power were adopted at first, with an idea that they would best supply a varying demand. The constructors of the station now think that the accumulator battery renders this unnecessary and that motors of a larger and uniform size would be more economical. They claim as advantages of a gas plant compared with a steam plant that less space and less water are required; that there is absence of smoke and danger of explosion; and that gas stations can be distributed more easily over the district to be supplied.

Distribution of Natural Gas at Pittsburgh, U.S.A.—A remarkable case of distribution of gas, for heating and power purposes, has been in operation at Pittsburgh.[2] The natural gas has almost entirely taken the place of coal in manufactories and for domestic heating, in a district where coal is exceedingly

---

[1] It is stated to be now 39 cubic feet per Kilowatt hour at the terminals of the dynamo.

[2] See a paper by Mr. Andrew Carnegie on ' Natural Gas,' read before the Iron and Steel Institute.

cheap. Coal can be obtained at Pittsburgh for 4s. to 5s. a ton, and coal slack at 2s. to 2s. 6d. a ton.

The natural gas was met with in boring for oil, and was first used to raise steam for the oil pumping engines. At 18 miles from Pittsburgh an enormous outburst of gas occurred, which for five years was allowed to burn to waste. Then a company engaged to take it a distance of 9 miles to Messrs. Carnegie's works. They were to be paid for the gas the value of its equivalent in coal until the capital cost of the pipes was repaid. After that the gas was to be supplied at half the cost of its equivalent in coal. In 18 months the cost of the pipes was repaid, and the gas was then supplied at half the cost of its equivalent in coal. It was afterwards conveyed into Pittsburgh and to still greater distances. When Mr. Carnegie described the operations, there were eleven gas mains, of 6 to 12 inches diameter, conveying gas to Pittsburgh.

The largest well discharged 30 million cubic feet per day, and other wells half that quantity. At the wells the gas had a pressure of 200 lbs. per sq. in., and at Messrs. Carnegie's works, 9 miles distant, the pressure was 75 lbs. per sq. in. This gave rise to difficulties from leakage, and it was found desirable to reduce the pressure in the pipes, in towns, and even to place ventilating pipes at every joint in the mains, leading the leakage above the level of street lamps. In using natural gas, one fireman can manage boilers developing 1,500 h.p.

Manufacture of Special Gas for Heating and Generating Power. A cheaper gas than lighting gas can be manufactured for heating and power purposes. (1) *Producer Gas*, obtained by forcing air through incandescent coke or anthracite. The resulting gas is mainly carbonic oxide and nitrogen. (2) So called *Water Gas*, obtained by injecting steam through incandescent coke or anthracite. Such gas has a volume of about 26 cubic feet to the pound, and will develop about 7,373 Th. U. per lb. In manufacturing water gas, air is first blown through the fuel till it is incandescent, and then steam, the alternation being repeated as necessary. (3) *Dowson Gas*, made by passing air and steam through incandescent coal or coke. This gas contains hydrogen, carbonic oxide, and a considerable quantity of nitrogen. Four volumes of it are about equal in calorific value to one volume of lighting gas. It develops 160 Th. U. per cubic foot, or about

2,382 Th. U. per pound. With anthracite at 13s. a ton, Dowson gas costs about 2d. per 1,000 c. ft. for fuel used, and exclusive of interest and depreciation of plant. Dr. Monaco gives the total cost of producing Dowson gas at 4d. per 1,000 c. ft., and its calorific power as one-fourth that of lighting gas. In that case Dowson gas is in heat value equivalent to lighting gas at 1s. 4d. per 1,000 c. ft. (4) Mr. Thwaite has proposed for power purposes a gas of about 12 candle power, obtained by mixing lighting gas and producer gas. According to him such a gas could be manufactured for 4d. per 1,000 c. ft., and distributed and sold at 1s. 4d. per 1,000 c. ft. Its calorific power is little less than that of lighting gas.

The following table contains data of the density and calorific value of various kinds of gas :—

DENSITY AND CALORIFIC VALUE OF GAS

	Cubic feet per lb.	Calorific value Th. U.		Oxygen required for combustion per c. ft. in c. ft.	Oxygen required for combustion per lb., in lbs.	Volume of products of combustion with air in c. ft. per c. ft.
		Per c. ft.	Per lb.			
Manchester gas				1·229	...	6·29
American gas	26·5	616	16,326	1·320	—	—
London gas .	33·7	617	20,801	1·157	3·18	—
Petroleum .	—		20,303	—	3·35	—
Pittsburgh gas	21·75	833	20,610	1·48	3·27	7·50
London gas .	30·3	633	19,199	1·24	3·35	—
Dowson gas .	11·89	160	2,382	0·24	0·32	2·71
Water gas (a)	26·0	284	7,373	—	—	—
,, ,, (b)	20·8	640	13,317	—	—	—
Dowson gas .	14·66	—	4,825	—	—	2·00

(a) Not carburetted   (b) Carburetted.

Formula for Flow of Gas in Pipes.—Let r_1, r_2 be the initial and terminal pressures in a main of length L (foot units). The velocity of flow is given by the equation [1]—

$$v_1 = \sqrt{\left(\frac{g \, c \, \mathrm{T} \, m}{\zeta \, \mathrm{L}} \quad \frac{r_1^2 - r_2^2}{r_1^2} \right)}$$

where v_1 is the velocity at the inlet of the pipe. For pipes of circular section and diameter d, $m = d/4$. For lighting gas

[1] Unwin; Distribution of Power by Compressed Air,' *Proc. Inst. C. E.*, vol. cv.

$c = 130$; for Dowson gas $c = 64$. Let the temperature be 60° F., or the absolute temperature $T = 521°$. Then $cT = 67,730$ for lighting gas, and $= 33,314$ for Dowson gas. ζ, the coefficient of friction, $= 0·003$. Introducing the numerical quantities, for lighting gas—

$$u_1 = \sqrt{\left\{181,700,000 \frac{d}{L} \frac{p_1^2 - p_2^2}{p_1^2}\right\}}$$

for Dowson gas—

$$u_1 = \sqrt{\left\{89,450,000 \frac{d}{L} \frac{p_1^2 - p_2^2}{p_1^2}\right\}}$$

where the pressures are in lbs. per square inch.

When the initial velocity of flow is given, and the terminal pressure is required in terms of the initial pressure, for lighting gas—

$$p_2 = p_1 \sqrt{\left\{1 - \frac{u_1^2 L}{181,700,000 \, d}\right\}}$$

for Dowson gas—

$$p_2 = p_1 \sqrt{\left\{1 - \frac{u_1^2 L}{89,450,000 \, d}\right\}}$$

CASE I.—In an ordinary gas distribution the difference of pressure producing flow is small, being about $2\frac{1}{2}$ inches of water. If $p_2 = 14·7$ lbs. per square inch, $p_1 = 14·7361$, and

$$\frac{p_1^2 - p_2^2}{p_1^2} = 0·00506.$$

The equations reduce to—

$$u_1 = 959·4 \sqrt{\frac{d}{L}}$$

for lighting gas, and

$$= 673·2 \sqrt{\frac{d}{L}}$$

for Dowson gas.

The quantity of gas delivered in cubic feet per hour will be

$$3,600 \times \frac{\pi}{4} d^2 u_1.$$

Assume a distribution to a distance of 5,000 feet. Then, with the given pressure difference of $2\frac{1}{2}$ inches of water column, the quantity of gas discharged and its equivalent in power will be as follows:—

POWER TRANSMITTED IN GAS MAINS

Length, 5,000 feet ; pressure producing flow, $2\frac{1}{3}$ inches of water column.

A.—Lighting Gas

Diameter of main in inches	Initial velocity in main, feet per second	Cubic feet of gas delivered per hour	H.P. at 26·5 cubic feet per h.p. hour
6	9·59	6,782	256
12	13·57	38,350	1,447
24	19·19	217,100	8,189
36	23·50	598,000	22,560

B.—Dowson Gas

Diameter of main in inches	Initial velocity in main, feet per second	Cubic feet of gas delivered per hour	H.P. at 90 cubic feet per h.p. hour
6	6·73	4,758	53
12	9·52	26,910	299
24	13·46	152,300	1,692
36	16·48	419,600	4,662

It will be seen that none of the velocities under this pressure are excessive. The Dowson gas being heavier, the friction is greater and the quantity flowing is less. Further, as the heat value of Dowson gas is less than that of lighting gas, the amount of power transmitted in a main of a given size is only about one-fifth as much for Dowson gas as for lighting gas.

CASE II.—It may next be inquired what would be the result of using greater pressure to force the gas through the mains than is usual in supply for lighting. In ordinary gas mains it is found unadvisable to increase the pressure, because of the increase of leakage. In the distribution of gas for power purposes this objection would have less weight. The network of mains would be simpler, and, the consumers being fewer, there would be fewer joints and valves to cause leakage. By the adoption of some really efficient joint, like that used in the Paris air mains, leakage could be almost reduced to zero. The pressure which would, then, seem to be desirable for a gas power distribution is the pressure which would produce in the mains the highest desirable velocity of flow. It may be taken from the analogy of compressed air mains that 45 feet per second is a quite unobjectionable velocity.

Assuming this velocity as the initial velocity in the mains, the problem is to find the necessary initial pressure. The equations become—

$$p_1 = \sqrt{\dfrac{p_2}{\left(1 - \dfrac{L}{89,730\,d}\right)}}$$

for lighting gas, and—

$$p_1 = \sqrt{\dfrac{p_2}{\left(1 - \dfrac{L}{41,180\,d}\right)}}$$

for Dowson gas.

For a transmission to a distance of 5,000 feet we get the following results:—

POWER TRANSMITTED IN GAS MAINS

Length, 5,000 feet ; p_2 = terminal pressure = 14·7 lbs. per square inch ; initial velocity, 45 feet per second.

Diameter of main in ins.	Initial velocity in ft. per sec.	Initial absolute pressure, lbs. per sq. in.	Difference of pressure producing flow in inches of water	Quantity of gas in c. ft. per hour	H.P. transmitted
A.—Ordinary Lighting Gas					
6	45	15·60	24·7	31,815	1,200
12	—	15·12	11·6	127,260	4,800
24	—	14·90	5·5	509,040	19,210
36	—	14·84	3·9	1,145,340	43,200
B.—Dowson Gas					
6	45	16·71	55·5	31,815	353
12	—	15·61	25·2	127,260	1,413
24	—	15·14	12·2	509,040	5,655
36	—	14·99	8·3	1,145,340	12,722

Cost of Gas Engines and Dowson Gas Plant.—The following estimate[1] of the cost of a gas plant for an electric lighting station may be useful for comparison with the cost of steam plant previously given.

Dowson gas plant for 1,160 i.h.p.	£2,000
Seven 122 b.h.p. gas engines	
Two 61 b.h.p. gas engines	9,012
Two 34 b.h.p. gas engines	
Dynamos and belting	5,495
For 696 Kilowatts, total	£16,507

[1] *Proc. Inst. C. E.*, vol. cxii. p. 95.

CHAPTER XIII

ELECTRICAL TRANSMISSION OF POWER

IN 1877, Dr. William Siemens indicated the practicability and the probable commercial importance of the electrical transmission of power to considerable distances. In an address to the Iron and Steel Institute he stated that a copper rod 3 inches in diameter would transmit 1,000 h.p. thirty miles. In 1883, he delivered a lecture at the Institution of Civil Engineers on ' The Electrical Transmission and Storage of Power,' but the only practically working transmissions which could then be described were the Lichterfelde and Portrush railways. About that time Marcel Deprez experimentally transmitted 3 h.p. a distance of 25 miles by ordinary telegraph wires, using a pressure of 2,000 volts, and obtaining at the motor only 32 per cent. of the energy expended.

It has from that time been hoped that the transmission and storage of energy for motive-power purposes would be one of the largest fields of electrical enterprise. Much progress has been made, especially in the last five years. But, having regard only to plants actually at work, it must be confessed that the total amount of power transmitted electrically and used for industrial purposes, exclusive of traction, is not yet very great. In Mr. Gisbert Kapp's Cantor Lectures on the ' Electric Transmission of Power,' only one plant of considerable magnitude, that at Schaffhausen, is described. At Schaffhausen two turbines, of 350 h.p. each, drive continuous current Oerlikon generators designed for an output of 330 ampères at 624 volts. Four cables, each of a section of 0·437 sq. in., convey the current 750 yards to actuate motors in a spinning mill. The power is sold at 2*l.* 16*s.* per h.p. per annum. This installation is to be considerably extended. In republishing his lectures Mr. Gisbert Kapp added a description of the Kriegstetten and Solothurn installation for transmitting 50 h.p. at a pressure of 2,000 volts a distance of 5 miles.[1] This is a continuous current system constructed by the Oerlikon Works. The commercial efficiency, when 23 h.p. were received at Solothurn, was 75 per cent. A number of other installations of a similar type, in which 50 to 300 h.p. have been transmitted 350 to 8,000 yards, have also been erected by the Oerlikon Company.

A very interesting continuous current transmission was constructed in 1889 by M. Hillairet, of Paris, at Domène, near Grenoble. A turbine of 300 h.p. drives a generator giving a current of 70 ampères at 2,850 volts. The current is conveyed 5 kilometres to actuate a motor in a paper mill. The efficiency of dynamo, line and motor is 65 per cent. The transmission has worked night and day with great regularity.

In all these cases direct currents were used. In 1891 alternating currents were employed in the striking Lauffen-Frankfort experiment. There 100 h.p., obtained from water power, was transmitted 108 miles with a loss of only 25 per cent. The experimental plant was erected by the co-operation of the Allgemeine Electricitäts Gesellschaft of Berlin and the Oerlikon Company of Zürich. High tension ranging from 16,000 to 30,000 volts was used in transmission. This was obtained by step-up transformers at Lauffen, and step-down transformers were used at Frankfort. The cost of the transmitting line has been stated at 15,000l.

The special object of the experiment was to illustrate a solution of the problem of transmitting power by alternating currents. With the three-phase system, motors which are self-starting, without commutators, and very simple in construction can be used. The Lauffen dynamo also was of simple construction and mechanically of excellent design. At 150 revolutions per minute it was capable of developing three alternating currents of 1,400 ampères each, at 50 volts above their common connection, equivalent to about 300 h.p. But it was usually worked at about 100 h.p. during the exhibition. The currents had 40 alternations per second. The conductors were of hard-drawn copper 4 mm. (0·16 inch) in diameter, having

[1] See also report by Professor Weber, *Die Leistung der Electrischen Arbeits-übertragung von Kriegstetten nach Solothurn*, Zürich, 1888.

a resistance of 2 ohms per mile. Each conductor served as a return to the other two. The line was carried on 3,227 wooden poles, the spans being 200 feet. The insulators were of porcelain, some of them having three oil grooves, but most only one oil groove. Both high and low tension circuits were grounded at the neutral point, or junction, of the three conductors. At Frankfort the current was transformed to 75 volts, and used partly for incandescent lamps, partly to drive a 100 h.p. motor. The greatest amount of energy transmitted is stated to have been 180 h.p.

As an example of yet another system of transmission, the installation by the Westinghouse Company at the Gold King mine, Telluride, Colorado, may be mentioned. There a Pelton water-wheel drives an alternating current generator. The current is carried by an overhead line on posts a distance of 3 miles, and actuates a synchronising motor of 100 h.p. The motor is started by a special Tesla motor.

In these transmissions, and in nearly all hitherto carried out, one or more generators drive one or more motors belonging to a single industrial undertaking. In such cases much of the difficulty and complication involved in a general distribution to many consumers are avoided, and the inconvenience and damage of a temporary stoppage, due to a breakdown of the line or lightning accident, are minimised. When the Niagara Commission met in London in 1891, only one case was known where power was distributed electrically to many consumers. That was the interesting installation at Oyonaz, not far from Geneva, erected by Messrs. Cuenod Sautter & Co. Turbines of 250 h.p. at Charmines generate a continuous current at 1,800 volts, which is transmitted 8 kilometres by an overhead line to Oyonaz. There the current is reduced in pressure by motor transformers, and is distributed, partly for lighting, partly for driving small motors in a number of workshops. With a supply of cheap power the village was very prosperous, when the author saw it in 1892 ; but at that time only about 30 h.p. was distributed for power purposes, and 40 h.p. for lighting.

In a great deal that has been said about the electrical distribution of power one thing has been too much overlooked. Means of transmitting power, even to considerable distances, have long been known. That they have not been more widely adopted is due, not to any risk of mechanical failure, but to the cost of transmission. If electrical transmission is to be extensively used, it must be when it can be carried out so cheaply, that power can be supplied at a less cost than that at which consumers can produce it for themselves.

Much has been accomplished in distributing electricity for lighting. But a higher price can be paid for electricity for lighting than for power purposes. Every Electric Lighting Company would be glad to supply current from its mains for power purposes, if only to increase the day load on the machinery and reduce the idle time. In Bradford, some electric motors are used for working hoists, lathes, &c. Recently, in London, electric motors supplied from the lighting mains have been applied to driving newspaper printing machinery. But the ordinary price of electricity for lighting purposes is 6d. per unit, which is equivalent to about 60l. per h.p. per year of 3,000 hours. At that price it can only be used for power purposes, either when the power is required for short periods intermittently, or where there is great local inconvenience in employing steam or gas engines. It is only where electricity costs from one-sixth to one-tenth of its ordinary price when used for lighting, that it can have any large importance as a means of obtaining power.

In the application of electricity to traction on tramways and town railways, a remarkable success has been achieved. In the United States there are from 4,000 to 5,000 miles of electric tramways, for which the power is distributed from power stations. But for traction as for lighting a high price can be paid for power.

For industrial purposes the question of cost is generally of controlling importance, and hence the progress of electrical methods of distribution has been less rapid. A review of the cases described above and others leads to the following conclusions as to the limitations of systems for distributing motive power electrically:—(a) When the power is initially steam power, its distribution electrically adds so much to its cost as to prohibit its transmission to any great distance in nearly all cases. (b) Hitherto it has only been in districts where cheap overhead conductors, carrying high pressure currents, can be safely used that electrical methods of transmission have proved commercially successful.

Conductors of Electricity.—The laws of electric flow are in many respects analogous to those of hydraulic flow. The resistance and the loss of pressure and energy depend on the length and section of the conductor. The amount of current which can be transmitted in any given case is limited by the heating effect, just as there are practical limitations to the velocity of flow in pipes. The permissible pressure in water mains depends on the strength of the pipes; the permissible electric pressure on the insulation.

A much wider range of electric pressure is allowable than can be permitted in fluid transmission. To carry current two miles at a given voltage, four times as much copper is required as to transmit it one mile, if the efficiency of transmission is the same. Hence long-distance transmission would be enormously costly but for the possibility of varying the voltage. By doubling the electric pressure in the two-mile transmission, the amount of copper required is reduced to the same as for the one-mile transmission. Hence the whole problem of cheaply transmitting power to great distances depends on the use of high electric pressures.

COMPARISON OF ELECTRIC CONDUCTORS

Material	Conductivity, pure copper = 100	Density	*a* Conductivity of equal weights, pure copper = 100	*b* Tenacity, tons per square inch	Product of *a* and *b*
Pure copper	100	8·9	100	17	1,700
Soft copper	98	8·9	98	13	1,280
Hard copper	97	8·9	97	29	2,810
Swedish iron	16·5	7·8	18·8	22·5	423
Galvanised iron	14	7·7	16·1	25	400
Cast steel	10·5	8·0	11·7	58	680
Aluminium	55	2·6	188	11	2,380
Silicon bronze	97	8·9	97	28	2,720
„ „	80	8·9	80	34	2,720
„ „	45	8·9	45	49	2,200
Phosphor bronze	26	8·9	26	45	1,170

Soft copper has the highest conductivity, and up to the present time it has been found to be the best material for conductors. Weight for weight, aluminium has nearly double the conductivity of copper, so that if its cost per lb. were less than half that of copper, which in time it may be, it would have almost equal merit. In the case of aerial lines, where in each span the conductor must sustain the stress due to its own weight,[1] the tenacity is of importance. The comparative value of conductors for aerial lines is about proportional to the product of the tenacity and conductivity of equal weights. Here again aluminium is twice as good as soft copper and nearly as good as hard copper or silicon bronze. Perhaps some alloy of aluminium may be found not much heavier, of equal conductivity, and of greater tenacity than the pure metal. In that case the alloy might have an advantage over any material at present used.

Laws of Steady Electrical Flow along Copper Conductors.— The weight of a copper conductor of a sq. in. section is,—

$$w = 3\cdot86 \; a \text{ lbs. per foot.}$$
$$= 20{,}380 \; a \text{ lbs. per mile.}$$

The electrical resistance per unit length is,—

$$r = \rho \, / \, a$$

where ρ is the specific resistance of the copper, which varies with its quality and temperature. For about $80°$ and copper of good quality, $\rho = 0\cdot0000086$ if l is in feet, and $\rho = 0\cdot045$ if l is in miles. A stranded conductor has about 28 per cent. more resistance, if a is understood to be the gross section, not deducting spaces between the strands.

Let I be the intensity of the current or quantity of electricity circulating in ampères per second, in a conductor of length l, section a, with a potential difference E at the ends. Then

$$E = r \, I \, l = \rho \, I \, l \, / \, a \text{ volts.}$$

The rate at which work is wasted in overcoming the resistance of the conductor is

$$\rho \, I^2 \, l \, / \, a \text{ watts}$$
$$= \rho \, I^2 \, l \, / \, 746 \; a \text{ horses power.}$$

When a generator drives a motor at a distance L there must be in general a going and return conductor, so that 2L must be substituted for l in the equations in finding the loss in the line.

Case of Alternating Currents.—When an alternating current is transmitted the current is not uniformly distributed over the

---

[1] Let w be the weight of the conductor in lbs. per foot run, l the span, and d the deflection in feet. Then the tension in the conductor due to its weight is—

$$T = 18 \, w \, l^2 / d \text{ lbs. per square inch.}$$

section of the conductor; there is a tendency to accumulation towards the surface. Consequently for large conductors and high frequencies the resistance increases disproportionately to the weight of copper used.

It is not possible to discuss the phenomenon here, but it is necessary to call attention to it as one of the special difficulties involved in the use of alternating currents.

Heating of the Conductor.—The energy wasted in the conductor is expended in heating it, and the heat is dissipated by radiation and convection. The conductor takes a temperature at which there is a balance between the heating and cooling. The rise of temperature which can be permitted is limited by the increase of resistance, the decrease of insulation, and possible injury or danger if the rise of temperature is excessive.

The most important experiments on the heating of conductors are those made by Mr. Kennelly in 1889. It is beyond the scope of the present treatise to consider these experiments in detail, and in the case of large conductors information is still defective. For cylindrical wires in still air the heating and cooling were found to be balanced when the following approximate equation was satisfied:

$$\frac{\rho \iota^2}{a} = (0 \cdot 073 \, d + 0 \cdot 029) \, t$$

where t is the rise of temperature of the conductor in degrees F. For blackened wires the cooling was twice as great. Professor G. Forbes has indicated that, for large conductors, thin strips will carry heavier currents than a solid conductor.

Efficiency of a Generator Line and Motor.—Let E_g be the potential difference at the terminals of the generator; E_m that at the terminals of the motor. Then, using the expression found for the line resistance,

$$E_g - E_m = \rho \iota l / a,$$

which determines the section of the conductor when the loss of pressure in the line is fixed.

If T is the electrical h.p. delivered by the generator, T_m that received by the motor,

$$T_g - T_m = \frac{\rho \iota^2 l}{746 \, a}.$$

The efficiency of transmission is

$$\eta_t = \frac{T_m}{T_g} = \frac{E_m}{E_g}.$$

If η_g, η_m are put for the efficiencies of the generator and motor, taking account both of electrical and frictional losses, the resultant efficiency of generator line and motor is

$$\eta = \eta_g \times \eta_t \times \eta_m.$$

Cost of Conductors.—In general, the method of transmission by aerial or underground, by bare or insulated conductors, will be decided by local or financial considerations. Then estimates can be made of the cost of conductors of assumed sections. Copper being an expensive material, a large part of the cost of the transmission (as distinguished from the generator and motor plant) is the cost of the copper and is proportional to the section of the conductors. There are some other charges which vary with the section of the conductors. But there are also charges which do not vary much with the size of the conductors. Hence the cost of the line of transmission per mile may be regarded as consisting of a constant part, and a part proportional to the section of the conductor.

Mr. Stuart Russell [1] has given some estimates of the cost of conductors used in electric lighting. From these, simplified a little, some of the following formulæ have been taken. Let a be the section of one conductor in square inches; L the distance of transmission in miles, so that for a going and return conductor the length of the conductors is 2 L: then the cost in pounds per mile of the line of transmission is:—.

	Cost per mile in pounds
Insulated cable in iron pipes or bitumen casing .	$897 + 4,118a$
Armoured cables in the ground	$792 + 4,435a$
Concrete culvert with bare conductors . . .	$1,584 + 1,795a$
Bare conductors on iron posts, with insulators for lines of large capacity	$500 + 1,609a$
Bare conductors on wooden posts, with insulators for light lines	$20 + 1,500a$

These formulæ give the cost of the line erected, but differences of cost of copper, labour and carriage may involve considerable differences of cost in particular cases.

[1] *Electric Light Cables.* London, 1892.

Condition determining the most Economical Section of Conductors.—Suppose current has to be conveyed from a generating station to a point of consumption at a distance L feet. In general the system to be adopted will be predetermined, and the cost of generating the current can be ascertained. Suppose, further, the kind of transmission selected so that the cost can be expressed in terms of the section of the conductor. If the fall of potential in the line is fixed, then the equation above gives the section of the conductor. The minimum section of conductor is fixed with reference to the heating permitted. Subject to this limitation, in other cases, financial considerations govern the size of the conductor. In a temporary installation the total capital cost of generator and line should be as small as possible. In some cases of generating by water power the total amount of current which can be generated is fixed, and then considerations as to the way in which the energy is to be disposed of may impose conditions on the amount of waste in the conductor. But most generally, in cases of extensive power transmissions, an indefinite amount of power is available at a fixed cost per h.p. at the generating station, and as much of it as is not wasted in the conductor can be disposed of at the end of the transmission, at a fixed charge per h.p. The consideration which then determines the most economical section of the conductor is this: the larger the conductor, the less waste of energy will occur in transmission, and the less will be the cost of producing that energy. On the other hand, the larger the conductor the greater will be that part of the cost of the transmission which depends on the section of the conductor, and the greater the annual charge on that part of the capital expended. Lord Kelvin pointed out, in 1881, that in such a case the most economical section of conductor is that for which the annual cost of the energy wasted in transmission is equal to the annual interest and depreciation on that part of the cost of the transmission which is proportional to the section of the conductors.

Case in which the Current is of Constant Intensity in Working Hours.—For simplicity suppose one generator driving one motor L miles distant, the total section of the conductors being $2a$. Let the capital cost of the line of conductors be expressed in the form

$$(a + \beta a)\,\text{L pounds}$$

where a and β are constants, some values of which are given above. If ε is the rate of interest and depreciation charged on the capital expended in pounds per pound,

$$\varepsilon\,(a + \beta\,a)\,\text{L pounds}$$

is the annual cost of the line of transmission. Now let g be the cost of producing one h.p. hour of electrical energy at the generating station, including interest on plant, maintenance, and working expenses; T the number of hours per annum during which energy is supplied. The cost g will be greater as the number of hours worked is less. From the equations above, the energy wasted in transmission will be

$$\frac{2\,\rho\,I^2\,L}{746\,a}\;\text{h.p.}$$

The annual cost of the energy wasted will be

$$\frac{2\,\rho\,I^2\,L\,g\,T}{746\,a}$$

The total annual expenditure in transmission will be

$$\varepsilon\,(a + \beta\,a)\,\text{L} + \frac{2\,\rho\,I^2\,L\,g\,T}{746\,a}$$

which is a minimum, if—

$$\varepsilon\beta\text{L} = \frac{2\,\rho\,I^2\,L\,g\,T}{746\,a^2}$$

$$\frac{I}{a} = \sqrt{\left\{ \frac{373\,\beta\,\varepsilon}{\rho\,g\,T} \right\}}$$

which gives the ampères per sq. in. of section of conductor. If the section is determined in accordance with this law, the annual cost of energy wasted is equal to the annual charge for interest and depreciation, on that part of the cost of the line of transmission which is proportional to the section of the conductor. The adoption of this density of current is subject to the condition that it does not cause excessive heating of the conductor. The current density is independent of the length of the transmission.

Case of a Varying Current.—The energy wasted in a given conductor is proportional to the square of the current. Suppose that in a given transmission the maximum intensity of the current is I, and that the current x_1 I is transmitted for t_1 hours; x_2 I for t_2 hours; and so on. Then the equivalent constant current, or constant current for which the waste of energy would be the same as that of the actual varying current, is

$$ \text{I} \sqrt{\left\{ \frac{(x_1{}^2t_1 + x_3{}^2t_3 + \ldots\ldots)}{\text{T}} \right\}} $$

where $\text{T} = t_1 + t_2 + \ldots$ the whole yearly hours of transmission. It is this equivalent current which is to be used in applying the equations above.

In the following calculations I is to be taken as the current intensity in each conductor, if the current is constant, or as the equivalent current intensity for the actual working hours, if the current varies.

Calculation of Proportions and Cost of Electrical Transmissions. In order to see how the cost of electrical transmission varies with local conditions of power available and distance, and with electrical conditions of voltage and system of transmission, a series of cases has been calculated. A simple transmission between generator and motor without complex distribution is assumed. The interest on cost of dynamo is supposed to be included in the cost of the power. That on the cost of motors is not taken into account, nor that on the cost of transformers, if required.

First of all, it is obvious that, speaking broadly, only those cases are suitable for electrical transmission in which the economical conditions governing the size of conductor can be complied with. In the following table the current density (ampères per sq. in. of conductor) has been calculated for various values of a h.p. year at the generating station, and for three assumed rates of interest on the capital cost of the line of transmission. The value of a h.p. year has been taken at 0·5*l*. to 10*l*. Taking the working year at 3,000 hours, this would correspond to a value per h.p. hour, for the power generated, ranging from $y = 0·04d$. to $y = 0·8d$. The cost of a mile of two conductors, excluding that part which does not vary with the section of the conductors, has been taken at $\beta = 1,795l$. when the two conductors have two sq. ins. of section.

274 DISTRIBUTION OF POWER

TABLE I.—CURRENT DENSITY

$$\beta = 1,795l.$$

$g\,T$ = cost of h.p. year at generating station in lbs.	I/a = current density in amperes per square inch, for rates of interest and depreciation		
	$\epsilon = \cdot 05$	$\epsilon = \cdot 075$	$\epsilon = \cdot 1$
0·5	1,220	1,493	1,725
1·0	860	1,050	1,220
2·0	610	745	860
4·0	430	525	610
6·0	350	430	495
8·0	305	374	431
10·0	273	334	386

The largest of these densities would probably involve excessive heating of the conductor. With these exceptions the numbers are practically suitable.

The following table gives the number of ampères in the circuit for different amounts of h.p. transmitted from the generating station, at various potential differences E_g at the terminals of the generator.

TABLE II.—INTENSITY OF CURRENT IN CIRCUIT

$$\text{H.P.} = E_g I/746.$$

H.P. transmitted	Current I in ampères for pressure in volts				
	100	500	2,500	5,000	10,000
100	746	149	30	15	7·5
500	3,730	746	149	75	38
1,000	7,460	1,492	298	149	75
5,000	37,300	7,460	1,492	746	373
10,000	74,600	14,920	2,984	1,492	746

For some of the cases given in this table, the following are the sections in square inches of each conductor of a line of transmission :—

TABLE III.—SECTION OF CONDUCTORS

$$a = I \sqrt{\frac{\rho g T}{373 \beta \epsilon}}$$

H.P. transmitted from generator	Economical section a of one conductor in sq. ins. for pressures at generator in volts				
	100	500	2,500	5,000	10,000

Case I.—Cost of line of transmission proportional to section of conductor = β = 1,600*l.* per mile; interest on cost of line $7\frac{1}{2}$ per cent. per annum (ϵ = ·075); cost of h.p. year at generating station = gT = 0·5*l.*

	100	500	2,500	5,000	10,000
100	·386	·077	·015	·008	·004
500	1·93	·386	·077	·039	·019
1,000	3·86	·772	·154	·077	·038
5,000	19·30	3·86	·772	·386	·193
10,000	38·62	7·72	1·544	·772	·386

Case II.—As above, but cost of h.p. year = gT = 2*l.* at generating station.

	100	500	2,500	5,000	10,000
100	·772	·154	·031	·015	·008
500	3·86	·772	·154	·077	·038
1,000	7·72	1·54	·309	·154	·077
5,000	38·60	7·72	1·544	·772	·386
10,000	77·20	15·44	3·088	1·544	·772

Case III.—As above, but cost of h.p. year = gT = 6*l.* at generating station.

	100	500	2,500	5,000	10,000
100	1·338	·267	·053	·027	·013
500	6·690	1·338	·262	·134	·067
1,000	13·38	2·676	·535	·268	·134
5,000	66·9	13·38	2·676	1·338	·669
10,000	133·8	26·76	5·352	2·676	1·338

The loss of pressure in the line in volts can now be calculated for the cases given in the last table.

$$\text{E} = \rho \text{l}_l' a = 2\rho \, \text{I} \, \text{L} / a$$

which, for a mile of transmission and for the value of $1/a$ given by Lord Kelvin's law, becomes

$$\text{E} = 2\rho \sqrt{\frac{373\beta\epsilon}{\rho g \text{T}}}$$

					Loss of pressure in volts per mile
Case I.		.		. .	127·0
„ II.		.		. .	63·5
„ III.		.		. .	36·6

т 2

Suppose it is assumed that, when more than 25 per cent. of the pressure at the generators is wasted in consequence of the resistance of the line, the conditions of transmission are unsuitable. Then some of the cases in the table above will be of this class. The conditions will be unsuitable if the voltage at the generator is less than

	Transmission 5 miles	20 miles
Case I.	2,540	10,160
„ II.	1,270	5,080
„ III.	732	2,928

Subject to this limitation, the amount of h.p. delivered at the motors can be calculated for the cases in the table above.

TABLE IV.—H.P. DELIVERED AT MOTORS

H.P. at generators	Five-mile transmission. Volts at generator					Twenty-mile transmission. Volts at generator				
	100	500	2,500	5,000	10,000	100	500	2,500	5,000	10,000
Case I.										
100	—	—	75	87	94	—	—	—	—	75
500	—	—	373	436	468	—	—	—	—	372
1,000	—	—	746	873	936	—	—	—	—	746
5,000	—	—	3,730	4,365	4,683	—	—	—	—	3,730
10,000	—	—	7,460	8,730	9,366	—	—	—	—	7,460
Case II.										
100	—	—	87	94	97	—	—	—	75	87
500	—	—	436	468	484	—	—	—	373	436
1,000	—	—	873	936	968	—	—	—	746	873
5,000	—	—	4,365	4,683	4,841	—	—	—	3,730	43,65
10,000	—	—	8,730	9,366	9,683	—	—	—	7,460	8,730
Case III.										
100	—	—	93	96	98	—	—	—	85	93
500	—	—	463	482	491	—	—	—	427	463
1,000	—	—	927	963	982	—	—	—	854	927
5,000	—	—	4,635	4,818	4,909	—	—	—	4,272	4,635
10,000	—	—	9,270	9,636	9,817	—	—	—	8,544	9,270

The following table gives the cost of line per mile of transmission in the cases already considered. For transmissions of 100 to 1,000 h.p. at the generating station, the cost is calculated by the formula, cost $= 20 + 1,600a$ pounds per mile, the conductors being supposed to be carried on wooden posts. For transmissions of 5,000 and 10,000 h.p., cost $= 500 + 1,600a$, the conductors being on iron posts. The distance of transmission does not affect the cost per mile.

TABLE V.—COST OF LINE PER MILE OF TRANSMISSION

H.P. at generator	Cost in pounds per mile for pressures in volts at the generators of		
	2,500	5,000	10,000
Case I.			
100	44	33	27
500	143	83	50
1,000	266	143	81
5,000	1,735	1,118	809
10,000	2,970	1,735	1,118
Case II.			
100	69	44	33
500	266	143	83
1,000	510	266	143
5,000	2,970	1,735	1,118
10,000	5,440	2,970	1,735
Case III.			
100	105	63	41
500	449	234	127
1,000	876	449	234
5,000	4,782	2,641	1,570
10,000	9,063	4,782	2,641

Finally, the cost of a h.p. year delivered at the motors can be calculated. The interest on the cost of the line of transmission is taken at $7\frac{1}{2}$ per cent. ; and the cost of a h.p. year at the generators is that given in the statement above, namely, Case I., $g\,T = 0\cdot5$; Case II., $g\,T = 2$; and Case III., $g\,T = 6$ pounds per year.

The following table is instructive, even if every allowance is made for the extent to which local circumstances, cost of carriage, and difficulties of various kinds in construction, may affect the cost per mile of a transmission. In the first place, it is clear how important high electrical pressure is in making economy of cost possible in long transmissions. Next, it is useful to note that, when high pressure can be adopted, the transmission adds so little to the cost of the power delivered that a greater expenditure is quite justifiable, if it secures more safety and security from accident. In high pressure transmissions greater expenditure on the line of transmission would not add so much to the cost of a h.p. at the motors that the use of the power would be seriously restricted. If high pressures are used, the simplest precaution to ensure safety against accident to life is to fence and patrol the line of transmission. This involves purchase of land and other expenses not included in the estimates given in Table VI.

TABLE VI.—COST OF A H.P. YEAR AT MOTORS

H.P. at generators	Five-mile transmission			Twenty-mile transmission		
	Cost in pounds of a h.p. year at motors for pressures in volts at generators of					
	2,500	5,000	10,000	5,000	10,000	—
Case I. —Cost of a h.p. year at generators 0·5l.						
100	·89	·72	·64	...	1 19	Wood posts
500	·81	·64	·57		·88	Wood posts
1,000	·80	·63	·56	—	·83	Wood posts
5,000	·84	·67	·60	—	1·00	Iron posts
10,000	·82	·65	·58	...	·89	Iron posts
Case II. Cost of a h.p. year at generators 2·0l.						
100	2·60	2·30	2·19	3·55	2·86	Wood posts
500	2 52	2·25	2·13	3·26	2·57	Wood posts
1,000	2·51	2 24	2·12	3·22	2·54	Wood posts
5,000	2·55	2·27	2·15	3·38	2·68	Iron posts
10,000	2·52	2·22	2 13	3·28	2 59	Iron posts
Case III. Cost of a h p. year at generators 6·0l.						
100	6·87	6·49	6·28	8·17	7·11	Wood posts
500	6 84	6·40	6·21	7·85	6·89	Wood posts
1,000	6·83	6·40	6·20	7·81	6·85	Wood posts
5,000	6·86	6·43	6 23	7·95	6·98	Iron posts
10,000	6·85	6 41	6·21	7 86	6·90	Iron posts

System of Electric Transmitting Mains.—For continuous currents the following arrangements are possible :—(*a*) Single conductor and earth return ; (*b*) Going and return conductor of equal size ; (*c*) Three-wire system, one being a balancing wire. The earth return is objectionable, partly because it interferes with telephone and telegraph systems, partly because it is dangerous to life, if the potential exceeds 500 volts. Method (*b*) is most suitable where there is one generator and one or more motors not distant from each other, and for distribution in series. In other cases method (*c*) has advantages. It would seem best to have the three conductors of equal size, so that, in case of accident to one conductor, the two others could be used to supply current, as in method (b). At the station where the three-wire transmission terminates, the distribution to local circuits can be from time to time re-arranged so as to keep the current in the balancing conductor small.

For alternating currents of single phase, a going and return conductor may be used. For alternating currents of two phases it is possible to transmit by three conductors, but it appears preferable to use four conductors of equal size. The three-phase system used in the Frankfort-Lauffen experiment requires three conductors of equal size, either acting as a return to the two others.

In all distributions for power purposes to a distance hitherto carried out, except some mining installations, bare conductors carried on wood posts have been used. In cases like Oyonaz, where the line is carried over fields to a small village, such a cheap method may be used without much objection, even when high pressure currents are transmitted. In that case, however, there is a liability to injury, especially to injury from frost and sleet damaging the insulation of the line, and to injury from lightning, which must be reckoned with. Malicious damage is also possible with aerial transmissions. In important aerial transmissions a patrol to detect and remedy defects seems necessary, and adds to the working cost. There are difficulties in using insulated cables for high pressures, besides their cost and the liability of the insulation to deteriorate.

It is fair to point out that, when the cheapness of electrical transmission is put forward as a reason for adopting it in preference to other methods of transmitting power, it is always assumed that such a rough expedient as overhead conductors on wooden posts can be adopted. To a mechanical engineer such arrangements do not appear to afford adequate security or permanence for an important power distribution. In proportion as the number of consumers taking power from a common source becomes greater, the inconvenience, cost, and damage of any temporary stoppage of the supply of power become more serious. Hence it will probably prove to be necessary, if any general system of distribution of power by electricity is carried out, to place the conductors in subways, where they are protected from injury. Such a construction, however, will necessarily increase the cost of electrical transmission.

The smallest self-respecting town requiring a water supply would not hesitate to build such a concrete conduit as that shown in fig. 74. D'Arcy built such a conduit 13 kilometres in length for the water supply to Dijon. It is the smallest conduit accessible throughout. An important electrical power distribution needs permanent and secure construction as much as a system of water supply. An objection is sometimes made to a subway for bare conductors carrying high pressure currents, that there would be danger to life in traversing the conduit. To obviate danger as much as possible the conductors have been placed in recesses. Further, by movable metal screens put in connection with a return or earthed conductor any part of the conduit could be made absolutely safe while repairs were in progress. The figure shows only a rough sketch of a possible arrangement, but some permanent protection for conductors will have to be adopted in important electrical distributions.

FIG. 74.

Electrical Systems.—The continuous current method of electrical working has hitherto been most frequently adopted. The alternating current produced in the dynamo is commutated into a continuous current for transmission, and commutated back into alternating currents in the motor. Direct current dynamos and motors are well understood. The motors are satisfactory, start with a load on, and have been largely used for tramway and other purposes. For transmission to moderate distances the direct current method is well adapted. On the other hand, it has two essential defects or limitations. The first is that the electric pressure of the direct current can only be altered by

expensive running machines termed motor transformers, or
motor and generator combined. If, for transmission economically,
a high pressure current is used, it can only be reduced in
pressure, for distribution in places where the high pressure is
dangerous, by motor transformers. Next, there appears to be a
limit to the electrical pressure which can be obtained in a
direct current system, due to the complicated construction and
difficulty of insulation of the commutators. It does not seem
practically possible to obtain more than 2,000 or 2,500 volts [1]
in direct current dynamos. By coupling these in series a higher
pressure can be obtained, but this involves new insulation difficulties, and probably 10,000 volts is the limit at which a continuous current system is likely to be worked.

There are two arrangements of direct current systems.
Ordinarily the current is transmitted from the generators at
constant pressure, and the motors are connected with the mains
in parallel. Motors can be connected to the mains or disconnected without interference with others, unless the line resistance is excessive. Working in parallel the pressure in the
mains is nearly constant, and the loss of energy in the line
diminishes as the load diminishes.

The second system of direct current working is to transmit
a constant current, with variable electric pressure, increasing as
the load increases, the motors being all in series. If all the
motors are disconnected, one dynamo is run at a voltage just
sufficient to send the full current through the mains. As motors
are added and the resistance increases, the dynamo is run faster
and the voltage increased. When one dynamo is working at
full speed, a second is run on closed circuit till it is delivering
the required current. It is then connected in series with the
dynamo already coupled to the mains. If the load still increases
the speed of the second dynamo is increased, and when it has
reached full speed a third dynamo may be added in the same
way. With this system the loss of energy in the mains is the
same at all loads. It is not, therefore, so efficient as the parallel
system, but it has in other respects advantages. The regulation
of the dynamos, so as to maintain a constant current by varying
their speed, is comparatively easy, and the regulation of the

[1] Pressures as high as 4,500 volts have been proposed in direct current
dynamos.

motors to constant speed by a centrifugal governor regulating
the excitation is also easy. For small motors the parallel
system is more convenient.

Messrs. Siemens Brothers & Co. proposed a series system of
this kind for the Niagara distribution in 1890, and Messrs.

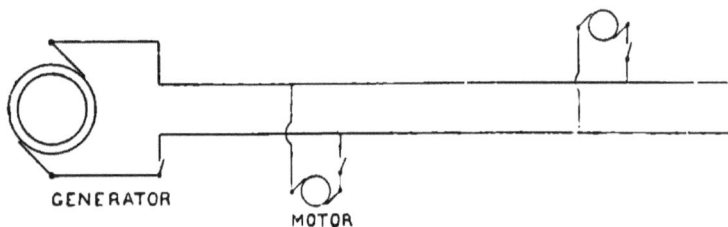

GENERATOR

MOTOR

FIG. 75.

Cuenod Sautter & Co. have since carried out a very important
power distribution at Genoa in the same way.

Fig. 75 shows an ordinary constant pressure, continuous
current distribution with motors in parallel. There is an
economy in copper, however, if a three-wire system is adopted.
For a long distance transmission it would appear best to have
the three conductors of equal section, so that any two could be

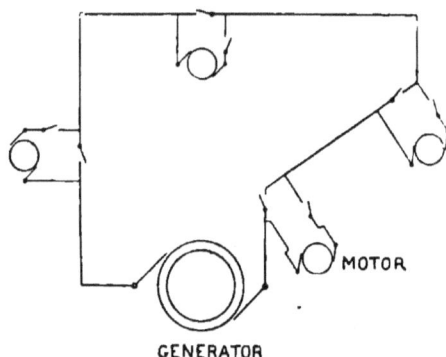

MOTOR

GENERATOR

FIG. 76.

used to transmit current in case of accident to one. Fig. 76
shows the constant current system of distribution with motors
in series.

During the last three years the alternating current method
of electrical working for power purposes has made considerable
progress. In this method the alternating currents produced in

the dynamo are transmitted without commutation. The insulation difficulty which limits the pressure in direct current dynamos is much less serious. Further, the electric pressure of an alternating current can be altered to a higher or lower pressure, with great facility, in inductive transformers having no moving parts, requiring no attention and easily insulated for almost any pressure. This facility of varying the electric pressure for different purposes is of enormous importance in a general system of electric distribution.

Up to a recent period alternating motors did not meet all requirements. An alternating dynamo generator will run as a motor, if supplied with current, synchronously with the generator supplying the current. But it has no starting torque, and requires to be put independently into motion at the right speed

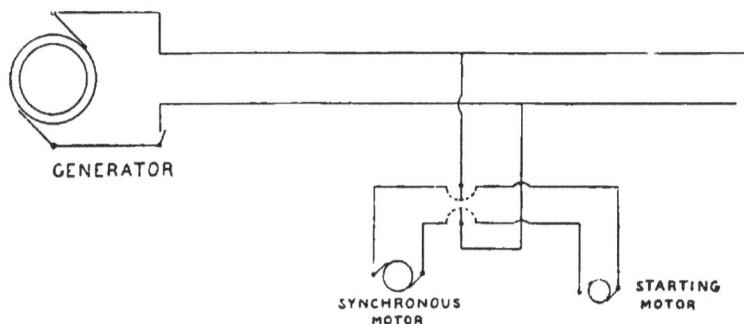

GENERATOR

SYNCHRONOUS MOTOR

STARTING MOTOR

FIG. 77.

before current is supplied to it. Once started in synchronism, it will keep step with the generator, if of suitable type, even if considerably over or under loaded. But the necessity of starting synchronous motors by an independent motor is for many purposes a serious defect. Next, the rotary field motors of Tesla and Ferraris were developed. For these two or more alternating currents differing in phase are required. These motors start with a load, and are now made both of a rotary field non-synchronous or of a synchronous type. More recently still motors have been constructed, which are self-starting, with an ordinary single-phase alternating current.

Fig. 77 shows the arrangement of an alternate current two-wire, or single-phase, transmission. The principal motor is a synchronising motor. But as this is not self-starting, a Tesla single-phase motor is used as a starting motor. The starting motor is first put into circuit. When this comes up to speed, it is used to drive the synchronising motor by a friction clutch. When the synchronising motor is at precisely the right speed, which is indicated by a special instrument (termed the synchroniser), the starting motor is disconnected, and the synchronising motor put into circuit. The operation can be carried out in two or three minutes.

The Westinghouse Company in the United States use in ordinary cases for transmissions of this kind a pressure of 3,000 volts and 120 alternations per second. The commercial efficiency of generators and motors varies from 88 per cent. in small sizes to 92 per cent. in large sizes. That is, a combined generator

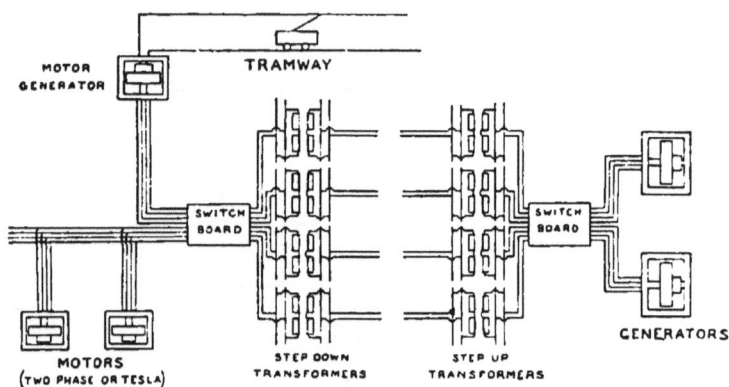

FIG. 78.

and motor would have an efficiency of 77 to 84 per cent., exclusive of loss in transmission. These efficiencies are at full load. At quarter load they are about 30 per cent. less.

Fig. 78 shows the arrangement of an alternate current polyphase system. Each generator delivers two currents to step-up transformers. These currents may be of low pressure, so as to be handled safely. After transmission, at a sub-station, the current is lowered in pressure by step-down transformers. The motors shown are two-phase Tesla motors which are self-starting. But a motor generator is also shown, producing a low pressure direct current for working a tramway. Three conductors could be used to each generator. but it appears to be preferable in long distance transmission to use four. Further, it

is now certain that alternate current dynamos can be so constructed that they can be worked in parallel, even when driven
by separate turbines or engines. In that case two or more
generators may deliver current into the same four conductors;
the dynamos once being run to the speed at which they fall into
step will not break step either with large variations of the
driving efforts, or the excitation, of the different generators.
With a two-phase system currents for lighting or for working
single-phase motors may be taken from a pair of conductors.

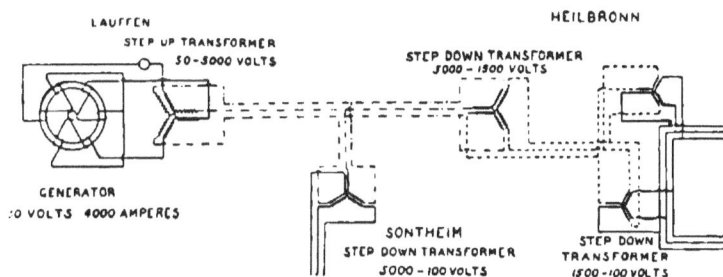

FIG. 79.

Fig. 79 shows the arrangement of the three-phase alternate
current system in use at Heilbronn. The 'Drehstrom' dynamo
produces current at 50 volts, which is transformed for transmission to 5,000 volts. This is transformed down in Heilbronn
in two stages to 1,500 and 100 volts. In the village of Sontheim
it is transformed directly to 100 volts. There are three equal
conductors carrying current in different phases, and all connected at a common neutral point.

CHAPTER XIV

EXAMPLES OF POWER TRANSMISSION BY ELECTRICAL
METHODS

IN general it will not pay to transmit power by electrical methods when the energy has to be generated by steam power. The exceptions are when special economies or conveniences result from electrical transmission, or where a high price can be paid for power, as in the case of tramways and town railways. One case where electrical methods have proved to possess some convenience is in distributing power to different workshops not widely distant from each other. The prime motor generates current which actuates secondary motors. conveniently located in the workshops, and thus cumbrous mechanical transmission is avoided. Generally, however, it is where water power is available as a source of energy that electric transmission of power can be most advantageously applied. The progress accomplished in this direction will be best explained by describing some typical cases.

It will be found that, for some time, direct current methods were alone used. But the need of higher electric pressures than are possible with the direct current system has compelled electrical engineers to study the construction of motors suitable for alternate currents. The synchronous motor has been well understood for some years, but there are obvious objections to the general use of a motor which can only be started by some independent motor. The three-phase motor of Brown and Dobrowolsky was the first alternate current motor which seemed to possess the practical advantages of continuous current motors. These advantages, however, are gained at the expense of complication in the line. The two-phase motors of Tesla have now been constructed in large sizes, and appear to meet most of the conditions of practical application. In small sizes, at any rate, single-phase, non-synchronous, alternate current motors have now, been constructed, by Brown, Tesla, and others, so that the initial difficulty of using alternate currents, that there were no trustworthy motors suitable for ordinary purposes, seems likely to disappear. The multiphase motors have the advantage over direct current motors that they have no commutators or brushes, and are simpler in design, while they have in common with them the ability to start against a load and to run non-synchronously, which is often an advantage. As to the present position of the question of the choice of an electrical system, Mr. A. Siemens said, in his recent address to the Institution of Electrical Engineers, that ‘ no hard and fast rule can be laid down that either the direct or alternate current system should be preferred in all cases, where power has to be transmitted to a distance. In judging of the merits of the two systems the proper conclusion can only be arrived at if the commercial aspect of the case is allowed to decide the question.’

CONTINUOUS CURRENT SYSTEMS

The Herstal Small Arms Factory in Belgium.—In 1866, the manufacturers of small arms in Belgium formed a syndicate, and, in order to carry out a large Government order, decided on the erection of new workshops of the most modern construction. M. Léon Castermans has given an account [1] of the development of the plans for these works, and of the reasons for the adoption of electrical transmission.

The operations to be carried on involved the construction of a number of different factories, so arranged as to be capable of future extension. In these factories it was found necessary to provide for 13 lines of shafting requiring a total of 200 h.p. or, allowing for loss in transmission and engine friction, about 300 i.h.p. Electric lighting was next decided on, and for this an additional 160 i.h.p. was needed.

It now became a question how these lines of shafting were to be driven from the steam engine. The mechanism between the steam engine and the lines of shafting in the workshops

[1] *Revue Universelle des Mines*, xix. 1892. There is an account of the Herstal Factory in the *Engineer*, November 25, 1892.

may conveniently be termed the intermediate transmission. This is required to sub-divide and distribute the motive power, and to effect necessary modifications of speed. The whole amount of power to be transmitted was not very great, and it had to be much sub-divided. If ordinary mechanism were used there would be a loss of power at each step of the process of distribution. The aggregate of these losses is large. It was a special inconvenience of a mechanical system of distribution that no part of the mechanism of transmission could be disengaged, except by the use of somewhat cumbrous appliances. Hence, practically, the whole transmission would be kept running even when part of the workshops were idle. The waste, considerable at full load, would be largely increased when only part of the machines were at work.

Two systems of transmission were first studied—shafting and gearing, and rope transmission. It was found that, for either system, there would be required some 30 tons of pedestals and fixed supports and 40 tons of moving mechanism. Practically, the whole of the moving pieces would be constantly running, whether much or little work was being done. There is a further disadvantage of such systems—they do not easily lend themselves to extensions of the works. Either the gearing must be initially of excessive size, or when an extension is necessary it must be removed and replaced by new gearing.

These considerations led to an investigation whether electric transmission could be adopted, with secondary motors driving each line of shafting.

Finally, it was decided to have one dynamo of 500 h.p., with its armature directly connected to the crank shaft of a steam engine ; to transmit the power electrically, and to have secondary electric motors for the lines of shafting and some special tools.

The mechanical efficiency of the engine has been found to be 94 per cent. The dynamo has an efficiency of 90 per cent., and the motors have an average efficiency of 87 per cent. at full load. Allowing two per cent. loss in leads, the resultant efficiency is $0.94 \times 0.87 \times 0.90 \times 0.98 = 0.72$; or 72 per cent. of the indicated power is delivered to the lines of shafting.

The electric system has the advantages—(1) simplicity in the transmission between the dynamo and secondary motors ; (2) saving of waste of work in consequence of the readiness with which any secondary motor can be disconnected; (3) great efficiency of transmission; (4) facility for future extensions with little modification of existing plant.

The installation has been very successfully carried out. The steam engine of 500 h.p. was built by the Société Anonyme, Van den Kerchove, of Ghent. It runs at 66 revs., and carries the armature of the dynamo on its crank shaft. The armature serves as a fly-wheel. The field-magnets are shunt wound. The dynamo can develop 2,400 ampères at 125 volts. There are two commutators, each taking half the current.

There is no doubt that, in factory driving, the distribution of secondary motors economises power. But it is fair to point out that the problem of distributing power by secondary motors in such cases did not first arise at Herstal, and can be solved with somewhat similar advantages by other methods. It has always been a question in large works, how far it is desirable to have a single steam engine, or smaller engines driving special parts of the works. In dockyards there has long been distribution of power to secondary motors by pressure water, or compressed air. In some works scattered gas engines have replaced a single steam engine, and much of the intermediate transmission is then dispensed with. For a long time at Seraing, and recently at some works in America, there has been a distribution of power by compressed air to secondary motors working special departments or machines.

Transmission for Mining Work in Nevada.—At the Comstock mines there existed a Pelton wheel of 200 h.p., on a fall of 460 feet. To obtain additional power, the water discharged from this wheel is conducted by two iron pipes down the vertical shaft and incline of the Chollar Mine, to the Sutro Tunnel level. It is there delivered under a head of 1,630 feet to six 40-inch Pelton wheels. Each wheel develops 125 h.p. at 900 revolutions per minute, the jet of water being only ⅞ of an inch in diameter. Each Pelton wheel is coupled to a Brush direct current dynamo. From these generators the current is taken to the surface, where it drives motors which work a 60-stamp mill. The efficiency of the electric apparatus is said to be 60 per cent., and the Pelton wheels weigh less than 2 lbs. per electric h.p.

Transmission at the Greenside Silver-lead Mines, Patterdale,

U

Westmoreland.—Water is supplied from a small tarn under Helvellyn through 15-inch pipes, and works a 100 h.p. vortex turbine under a head of 400 feet. This drives a four-pole compound direct current dynamo. The current at 600 volts is taken 6 furlongs by bare conductors on poles, and thence into the mine by insulated cables. In the mine there is a 9-h.p. series motor for winding, a 9-h.p. motor for pumping, and a motor transformer giving a current at 250 volts, which works motors on a tramway. There is also some lighting by incandescent lamps in series of six.

The Genoa Installation.—Exceptional circumstances have made it possible, at Genoa, to establish an electric supply in connection with a water supply. Some ten or twelve years ago, works for supplying water to Genoa additional to others then in operation were constructed. The water is obtained from streams on the Piedmont-Liguria frontier and impounded in reservoirs on the Gorzente River, an affluent of the Po. The reservoirs are 2,050 feet above Genoa, there being a large surplus fall not required as head for the water supply. To relieve the pipes of unnecessary pressure three relieving or service reservoirs were constructed, reducing to about 600 feet the pressure available for the transmission of the water to Genoa. The first reservoir is near the outlet of a tunnel by which the water is carried through a mountain ridge, and is 360 feet below the tunnel. At that point 730 gross h.p. can be utilised. The second reservoir is 360 feet below the first, and there also 730 gross h.p. can be obtained. The third reservoir is 500 feet below the second, and there 1,000 gross h.p. can be obtained. The water supply scheme has not been entirely successful, and the engineer, M. Bruno, and the consulting engineer, M. Prève, were led to consider the utilisation of the surplus fall to generate electricity to be transmitted for lighting and power purposes to Genoa, distant about sixteen miles.

At the three points described electric generating stations, named after the Italian electricians Galvani, Volta, and Paccinotti, are now in operation.

A first installation of a turbine of 140 h.p. was made in 1889 at the Galvani Station. This proved successful, and the further development of the stations was undertaken. The

Galvani Station supplies electricity to fifteen motors distributed along the valley from Isov orde to Genoa, and to a motor of 60 h.p. at the railway station in Genoa. It also supplies motor transformers at the Central Electric Lighting Station in Genoa. The remainder of the power at this station, amounting to 600 h.p., is utilised by means of telodynamic transmission. The Volta Station supplies electricity for lighting the station of Sempierdarena, by a motor of 60 h.p. driving twelve Siemens and two Technomasio dynamos, and also supplies electricity for motive power to a number of mills and factories, and the repairing shops of the railway. More recently the Paccinotti Station has been put in operation, and there are now two circuits, one fed from the Volta, the other from the Paccinotti Station. Messrs. Cuenod Sautter & Co., of Geneva, who installed the electrical plant, have furnished the following details. Messrs. Rieter Brothers, of Winterthur, constructed the first turbines erected. The remainder have been constructed by Messrs. Faesch & Piccard, of Geneva.

Electrical System.—The distribution is in series at constant current. The generating dynamos maintain a current of constant intensity in a single circuit which traverses all the motors. They supply a constant number of ampères, whatever the number of motors at work. The voltage is essentially variable. At certain hours all the motors are out of action; then the dynamos furnish a current at 450 or 500 volts, corresponding to the loss in the circuit. At certain hours both the motors supplying power and the motors driving dynamos for lighting are in action; then the voltage reaches 5,000 to 6,000 volts.

Galvani Station.—This has a single group of machines, consisting of two Thury continuous current dynamos connected by Raffard couplings to a Rieter turbine of 140 h.p., having a normal speed of 450 revs. per minute. In addition, a jute factory absorbs the power of two Rieter turbines of 300 h.p., the power being transmitted to it by telodynamic cables. The generators have six poles and give at full load 47 ampères at 1,000 to 1,100 volts. Their speed varies from 20 to 475 revs. per minute, according to the demand for power. The dynamos are coupled in series and work day and night.

Volta Station.—This has been in operation since 1891. There are four turbines of 140 h.p. each, and eight dynamos working at 47 ampères and 1,000 volts at full load. These generators work at constant speed, and the regulation is effected by varying the exciting current. The regulation is effected by a single regulator, however many generators are in action. The main turbines are Faesch & Piccard turbines, with relay governors which maintain a strictly constant speed.

The regulation of the exciting current involves difficulties, because the voltage of the main circuit varies from moment to moment, and sometimes quite suddenly, when motors driven by the current are thrown out of action. The motors are thrown out instantly by short-circuiting them. To meet these conditions the exciting dynamo is driven by a separate 15 h.p. turbine, which has as little inertia as possible. The exciting dynamo has a very light armature, so that it follows instantly the variations of speed of the turbine driving it. The exciting dynamo is itself excited by a small machine serving to light the station, and the stability of its magnetic field is thus independently secured. The turbine driving the exciting machine is provided with a relay governor, but the conical pendulum ordinarily used to secure constant speed is replaced by a solenoid, holding in equilibrium a soft iron core weighing 33 lbs., which is directly attached to the valve of the relay. A leather belt keeps this core and the valve in rotation, so as to practically annul any frictional effect. A spring and counterweight permit the adjustment of the action of the regulator. The solenoid is traversed by the current in the main circuit, which is normally 47 ampères. If the current augments, the core of the solenoid rises, puts in action the relay, and closes the sluices of the turbine driving the exciter. Vice versâ, if the current decreases the relay opens the turbine sluices.

The Paccinotti Station.—The system at the Volta Station has given good results, but at the Paccinotti Station a return has been made to a system similar to that at the Galvani Station. This station has four groups, each consisting of one turbine of 110 h.p. and two dynamos. The turbines have no speed governors or fly-wheels. The dynamos are self-exciting, with very light moving parts. The regulation of the current is effected thus. The turbine sluices are actuated by Piccard relay motors. The slide valves of the four relay motors can be separately connected to or disconnected from a shaft running

the whole length of the building.   This shaft is driven by a
1 h.p. electric motor, which has on its armature two windings
in opposite directions.   Each winding has its own commutator.   The problem is then to act on the motor so that it shall
revolve in a direction causing a closing of the turbine sluices
when the current increases, and an opening of them if the
current diminishes.   To obtain this there is an apparatus acting
as a sensitive ammeter.   It consists of a relay acted on by the
main current, the movable tongue of this relay, controlled by a
spring, being in equilibrium with the normal current.   If the
intensity of the main current varies above or below the normal
the relay tongue is attracted in one direction or the other.
According to the direction in which the relay tongue moves
current is sent through one or other of the armature windings
of the 1 h.p. motor, and the Piccard relays open or close the
turbine sluices.   As the regulation acts on the four turbines
simultaneously, a very small movement compensates for large
variations of current.

On account of the lightness of the moving parts of the
turbines and dynamos, the action is in effect very like automatic
regulation.   When the resistance of the circuit is increased by
the addition of motors the current diminishes, the torque also
diminishes, and the speed of the turbines increases, thus tending
to bring back the current to its normal value.   The reverse
effect occurs if the current increases.   The motor and relay
regulators have thus only to correct small variations.

The Transmitting Arrangements.—The current is transmitted
by bare copper wires three-eighths of an inch in diameter,
carried on posts, with porcelain and oil insulators.   The
greatest distance of transmission is thirty miles.   Each circuit has
a resistance of about 500 volts, so that the line efficiency at full
load is about 90 per cent.

The Motors.—The motors are all Thury motors, and are from
5 to 60 h.p. From 5 to 18 h.p. they are bipolar. Their
regulation is effected by shunting more or less of the current,
and they are all placed in series. Larger motors are multipolar,
and these are regulated by displacing more or less the points at
which the current enters and leaves the motor. This has the
effect of reducing the magnetic field, by causing some of the
convolutions to be traversed in a direction reverse to the

normal. Each motor is governed by a relay, and has a fly-wheel. A lever and counterweight is provided to adjust the governor.

The ratio of the effective work at the motors to the effective work of the turbines is stated to be 72 per cent.

Transmission at Biberist, near Soleure.—Messrs. Cuenod Sautter & Co. have also carried out a power transmission between Ronchatel and Biberist. Turbines of 360 h.p. have been erected, and the power is transmitted 28 kilometres to a paper mill. The generating station has two Thury continuous current dynamos, coupled to the turbines by Raffard couplings, and running at 275 revs. per minute. Each dynamo gives a current of 3,300 volts, and the two dynamos are coupled in series. The current is transmitted by a bare copper wire, $\frac{1}{4}$ inch in diameter, on simple porcelain insulators. The receiving station has two dynamos similar to the generators, making 200 revs. per minute, and driving directly by Raffard couplings the machinery of the mill. An efficiency of 70 per cent. is guaranteed.

ALTERNATING CURRENT SYSTEMS

The Lauffen-Heilbronn Transmission.—This transmission is interesting as an example of the 'Drehstrom' or three-phase method of transmission, and because an effort has been made to combine power supply with light supply. The owners of the Würtemburg Cement Works at Lauffen, on the Neckar, having surplus water power, conceived, in 1889, the idea of utilising it to supply electricity to Heilbronn, some six miles distant. They accepted plans for using the three-phase current with step-up and step-down transformers. The generating dynamo at Lauffen gives 4,000 ampères at 50 volts. This is transformed to 5,000 volts for transmission, and then reduced to 1,500 volts in Heilbronn for distribution. Part is further reduced by second-ary transformers to 100 volts in a network for lighting. The charge for current used for lighting is 9*d.* per kilowatt hour. Current used for power purposes is charged at 4*d.* per kilowatt hour, or about 42*l.* per effective h.p. year of 3,000 working hours. In November 1892, there were on the system only 11 motors, aggregating 32 h.p., and it is believed that the use of electricity for power purposes is not even now very considerable. To prevent sudden fluctuation of the lighting circuits due to switching in or out of the motors, fluid resistances are arranged so that full contact cannot be made in less than 15 to 20 seconds. Motors over 3 h.p. are coupled direct to the high pressure mains. It is curious that as the energy supplied by the water is costless, it has been found economical to use surplus current in times of light load for drying the clay used in the Cement Works.

The New Electrical Scheme at Geneva.—The hydraulic installation at Geneva has already been described. The total amount of power available at Geneva will shortly be completely utilised. To obtain a further supply of power, a site has been found at 6 kilometres below Geneva (8 kilometres by the river) where a fall of 16 feet is available in high water, and 27 feet in low water, of the Rhône. The total amount of power available is 14,000 h.p. It is proposed to utilise the power at this site by 15 turbines of 800 h.p. each. The works are already in progress. For transmission the electrical method is to be used, and there is no doubt that alternating currents will be employed.

Transmission of Power at the Gold King Mine, Telluride, Colorado.—This installation is interesting as one of the first applications of alternating currents to the transmission of power.[1] A Pelton wheel on a fall of 320 feet drives an alternate current generator at 833 revolutions per minute, giving 166 alternations per second. The current is carried by bare conductors on posts to a crushing mill three miles distant, where it drives a synchronizing motor of 100 h.p. at 3,000 volts. The generator has a composite field winding. Part of the field magnets are excited by direct current from a separate exciter. The rest are excited by a current from the armature, proportional to the main current, and commutated by the equivalent of a two-part commutator. The pressure at the terminals of the generator rises as the current increases, so as to compensate for the increase of line loss, and to keep the pressure at the motor constant. Generally no adjustment is required after the motor is started. The motor is similar to the generator, but the field current is obtained from a second winding parallel with the

[1] See 'Long Distance Transmission for Lighting and Power,' by C. F. Scott, *Trans. Am. Soc. of Electrical Engineers*, 1892.

main coils on the armature. This current is commutated and passes through the field coils. To start the synchronizing motor a Tesla rotating field motor is used. This motor comes quickly to its normal speed, and then is used to bring the large motor up to speed. Both motors are belted to a countershaft, so that at the normal speed of the Tesla motor the synchronous motor is running above its normal speed. The large motor field magnets are then excited, and it runs as a self-exciting generator. The Tesla motor is switched off and the speed of the synchronous motor begins to fall off. At the moment when it has the same speed and phase relation as the distant generator (as shown by lamps), the main current is switched on. The small motor is disengaged by a friction clutch, and the load put on the synchronous motor by another clutch. Full load may be thrown on suddenly. The plant was started in June 1891. The aggregate time lost in the first nine months' working, from mishaps, was 48 hours. The plant has worked so well that it is to be increased by erecting a 750 h.p. generator and other motors.

CHAPTER XV

THE UTILISATION OF NIAGARA FALLS

FEW persons can have seen Niagara Falls without reflecting on the enormous energy which is there continuously expended, or without some trace of a feeling of regret that a supply of motive power of such enormous commercial importance should be wasted. To any engineer it must have occurred that the constancy of volume of flow and small variation of levels, the height of the fall, the suitability of the rocks for engineering operations, the proximity of railways and access by water to the great lakes, all marked out Niagara as an ideally perfect site for a manufacturing centre with factories worked by water power.

A vast system of inland lakes extends halfway across the continent of North America, on the boundary between Canada and the United States. The lakes form storage reservoirs for the rainfall on an area of 240,000 sq. miles, and they discharge at an almost uniform rate the collected drainage through the St. Lawrence to the Atlantic Ocean. Between Lakes Erie and Ontario, the last of the chain, the water flows through the Niagara River, which falls 326 feet in a distance of 36 miles. The average discharge through this short river has been estimated at 275,000 to 300,000 cubic feet per second. Such a volume of flow on such a fall would yield, if it could all be utilised, seven million horses power.

The Earlier Works for Utilising Niagara.—From a very early time water mills were built in the rapids above the falls. These older mills, which greatly disfigured the appearance of the falls, have now been cleared away. In 1853, a more important scheme was started, and by 1861 the so-called hydraulic canal had been constructed. This canal, originally 35 feet wide and 8 feet deep, receives water from the upper river at Port Day,

nearly a mile above the falls. It conveys the water to a line of
mills ranged along the bluff above the lower river. Turbines
in these mills discharge their tail-water on the face of the cliff
below the suspension bridge. By means of this old canal about
6,000 h.p. has been utilised, during the last quarter of a century.
The mills have been prosperous, and have been an important
factor in the industrial life of the district. They employ more
than a thousand operatives, to whom 70,000*l.* a year is paid in
wages. Recently. stimulated by the progress of another much
larger scheme. the old hydraulic canal has been enlarged, and is
now capable, it is said, of supplying turbines of 30,000 h.p.

   The Origin of the New Scheme.—About eight or ten years
ago, the governments of the Province of Ontario and the State
of New York began to be ashamed of the disfigurement of the
falls, partly by mills and still more by ugly buildings erected by
enterprising tradesmen. To restore and preserve the natural
beauty of the falls, land was purchased and formed into State
Reservations. within which objectionable buildings and other
disfigurements were removed. Then arose a feeling that no
more power could be obtained in the old way by mills along the
bluff. The idea of a better method of utilising the falls is due
to Mr. Thomas Evershed. He proposed to construct canals on
unoccupied land a mile and a half above the falls. From these
canals the water was to be supplied to turbines in vertical pits.
The tail-water from the wheel pits was to be collected by underground tunnels into a great main tail-race tunnel, passing under
the town of Niagara Falls and discharging into the lower river.

   In 1886, the Niagara Falls Power Company was formed to
carry out Mr. Evershed's plans, and a charter was obtained from
the State of New York, giving full rights to construct the
works necessary to utilise 200,000 h.p. and to convey the power
through any civil division of the State. A tract of 1,500 acres
of land was acquired, about 1½ miles above the falls. A subordinate company, the Cataract Construction Company, was constituted to carry out the necessary engineering works. and under
its direction a great tail-race tunnel, 7,000 feet in length and
capable of discharging the tail-water of turbines of 100,000 h.p.,
was commenced and has since been completed. The company
have the necessary powers to construct a second tunnel whenever
it is required.

When Mr. Evershed's plan came to be considered in detail, it appeared that the removal of the mill sites up-stream and the discharge of the tail-water underground solved only part of the problem. Mr. Evershed contemplated only the supply of water to the mills and its removal afterwards, as in some other American water-power works. The mill-owners were to sink wheel-pits and erect the turbines. In part this plan is actually being carried out, but it is only suitable for factories requiring large amounts of power. The growth of an industrial city supplied with power in this way would necessarily be slow, and the capital sunk in the construction of the great tunnel and other works would for years yield no adequate return. The network of surface canals would be expensive and the network of underground tunnels still more costly.

Obviously, it would economise the expenditure to develop a great part of the energy in power stations, by turbines of large size, aggregated in one place and placed under common management. It would facilitate the sale of the power, if manufacturers could have it delivered to them in any quantity required without trouble or expense on their part. More important still, if once power distribution instead of water distribution was accomplished, there would be markets for the power in towns at a considerable distance from Niagara. With so large an amount of power available, it was of enormous importance to obtain access to markets where motive power could be disposed of. The more the problem of power transmission has been studied, the more probable it has become that a great part of the power developed at Niagara will be used in towns at a considerable distance. Provisional arrangements have already been made, with a view of distributing power from Niagara in Albany, Rochester, Syracuse, and Schenectady, and for working the navigation of the Erie Canal, provided present expectations as to the commercial possibility of transmission to great distances are proved to be well founded.

The International Niagara Commission.—In 1890, Mr. E. D. Adams, the President of the Cataract Construction Company, and Dr. Coleman Sellers visited Europe. It was during this visit, and chiefly as a result of examining the works for distributing power in Switzerland, that the importance of the problem of distributing power instead of water was first clearly realised. It was resolved to invite selected engineering firms to send in completely worked out projects for the development and the transmission of power at Niagara, with estimates of cost.   To secure a perfectly unbiassed consideration of these projects, a commission was formed consisting of Lord Kelvin, Professor E. Mascart of Paris, Colonel Turrettini of Geneva, and Dr. Coleman Sellers of Philadelphia.   A sum of 4,500l. was placed in the hands of the Commission, to be awarded, partly in premiums to all competitors who sent in projects complying with the conditions, partly in prizes to those projects judged to be meritorious.   Plans and instructions were prepared for the information of competitors. In the event of any project being adopted, the conditions of payment were specified, both in the event of the machinery being constructed abroad and in the event of its being constructed in America under the superintendence of the competitor.   A number of projects were received, some for the utilisation of the water only, some for the transmission and distribution of the power, and some both for the hydraulic and transmitting machinery. The more important and fully considered projects were fourteen in number ; and amongst these, in one distribution by wire ropes was proposed, in two distribution by compressed air, and in one distribution by pressure water.   In all the other important projects electrical distribution was assumed ; Messrs. Cuenod Sautter & Co., Messrs. Vigreux and Levy, and Messrs. Hillairet and Bouvier proposed continuous current working at constant pressure ; Messrs. Siemens proposed continuous current series working at constant current and varying potential ; Messrs. Ganz and Professor G. Forbes proposed alternate current working.   The electrical plans of Messrs. Ganz and Messrs. Siemens were not worked out as detailed projects.   The Oerlikon Company were to have sent in plans for alternate current distribution, to accompany the hydraulic plans of Messrs. Escher Wyss & Co., of Zürich ; but they were not ready, and withdrew at the last moment.

Touching as briefly as possible on matters now chiefly of historical interest, it may be stated that many of the competitors appeared before the Commission and explained and discussed their plans.   It was assumed at that time that the greater part of the power would be distributed within two miles of the turbines, but that 25,000 or 50,000 h.p. might probably be sent to Buffalo, eighteen miles distant.   This should be remembered in considering the conclusions of the Commission. The project of Professor Riedler showed that it was probable that power could be transmitted to Buffalo by compressed air, at a cost less than that of steam power produced in Buffalo. But the Commission decided in favour of transmission by continuous current electrical methods. They were of opinion that no one of the projects submitted could be carried out without modification. Amongst questions arising out of the plans submitted they decided in favour of placing the electric machinery above ground, and not in underground rock chambers. They also suggested that to secure safety and freedom from accident, the electric conductors should be placed in covered subways.

The project at Niagara involves special difficulty in two ways. The works are of unprecedented magnitude, not only in the aggregate, but in detail and in the size of individual units. Also the commercial condition is an absolutely governing one. Steam power is cheap in the district, and unless the water power can be distributed at a price less than that of steam power, the market for it would be very restricted. The fall is so advantageous that the cost of power at the turbine shaft will be extremely low. But if heavy charges are added to this due to the expenditure on distributing machinery, its development will be useless. Many plans which have been proposed are impracticable, not because they would fail to work mechanically and electrically, but because they are too costly.

The Scheme now being carried out on the United States side.— The works now being carried out, and in part completed, comprise—

(1) The construction of a head-race canal 200 feet wide, 1,500 feet long, and 17 feet deep. This is completed.

(2) The construction of the great tunnel tail-race 6,700 feet long, 21 feet high, 18 feet 10 inches wide, with a slope of from 4 to 7 feet per thousand. The net section of the tunnel is 386 square feet, and it will discharge the tail-water of turbines developing 100,000 h.p. The velocity of the water in the tunnel when in full operation will be 18 to 25 feet per second. The floor of the tunnel, at the point of discharge, is 205 feet below the cill of the entrance sluices at the upper river, giving an effective fall of 140 feet at the turbines. It was hoped that the rock was so strong that lining would be unnecessary. This proved not to

be the case, and the tunnel is lined with four rings of brick in cement. For 200 feet back from the discharge portal it is lined with steel plates backed with concrete. Taking the discharge at 10,000 cubic feet per second, it amounts to about 3½ per cent. of the flow of the Niagara River.

(3) Works for the distribution of water to large consumers who erect their own machinery, with lateral tunnels to discharge the water into the tail-race tunnel, have been commenced. The first enterprise which took advantage of this arrangement was the Niagara Falls Paper Company (fig. 80). This company contracted to take 3,300 h.p. at once, with a right to take as much more subsequently. The mill has been built, and is said to be the largest paper mill in the world. The wheel-pit has been sunk 167 feet deep, and 28 feet by 43 feet in section. A vertical supply pipe, 13½ feet in diameter, conveys water to three Geyelin, inverted Jonval turbines of 1,100 h.p. each. The tail-water flows away, by a tunnel 7 feet diameter and 687 feet long, to the main tunnel. The power of the turbines is transmitted to the ground level by 10-inch vertical shafts; these drive the mill shafting by bevil wheels, which have 20 inches width of face and 6 inches pitch. The upward pressure of the water on the turbine wheels balances the weight of the vertical shafts. The shafts run at 260 revolutions per minute. The three turbines were recently started with success, and the mill is now in operation.

The Paper Company get their land and water power (exclusive of the cost of the machinery) at $8, or about 35s. per h.p. per annum, having the right to use the power 24 hours per day. The Pittsburgh Reduction Company, who control patents for the production of aluminium electrically, have acquired a site for works and the right to use 6,000 h.p. Other undertakings also are negotiating for sites and water supply under similar conditions. These works are carried out on the plan originally intended by Mr. Evershed.

(4) A large power station has been commenced, from which energy will be transmitted electrically both to consumers at Niagara and to Buffalo, and perhaps to towns more distant. The power house, shown in fig. 80, is placed alongside the head race canal, and is designed to contain 10 turbines of 5,000 h.p. each. Three of these have already been constructed. A wheel-

pit or slot has been sunk, which at present is 110 feet in length, 21 feet in width, and 178 feet in depth, having room for four turbines. It will be extended in length as required. Over this the first section of the power house has been erected, and it contains a 50-ton electric travelling crane, commanding the whole floor, which is being used in erecting the machinery.

The 5,000 h.p. Turbines.—Soon after the meeting of the Commission it was decided that the first turbines for the power

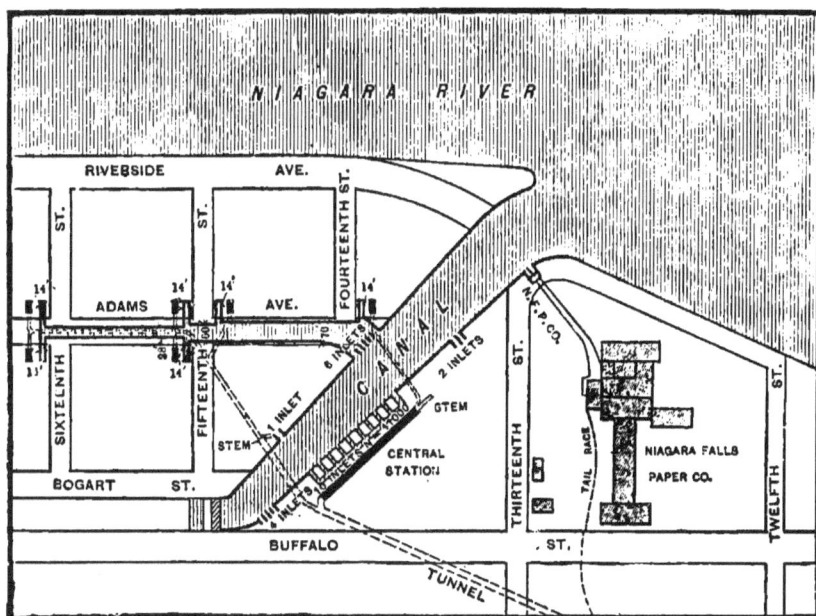

FIG. 80.

station should be of 5,000 h.p., running at 250 revolutions per minute, to suit the requirements of the electrical engineers. It was decided to place the turbines at the bottom of an open slot or wheel-pit, 175 feet in depth, and transmit the power to the dynamos at the ground level by vertical shafts. Designs by Messrs. Faesch & Piccard, who conjointly with Messrs. Cuenod Sautter & Co. sent in the combined project which received the highest award of the Commission, were selected. The first three turbines have been constructed by Messrs. J. P. Morris & Co., of

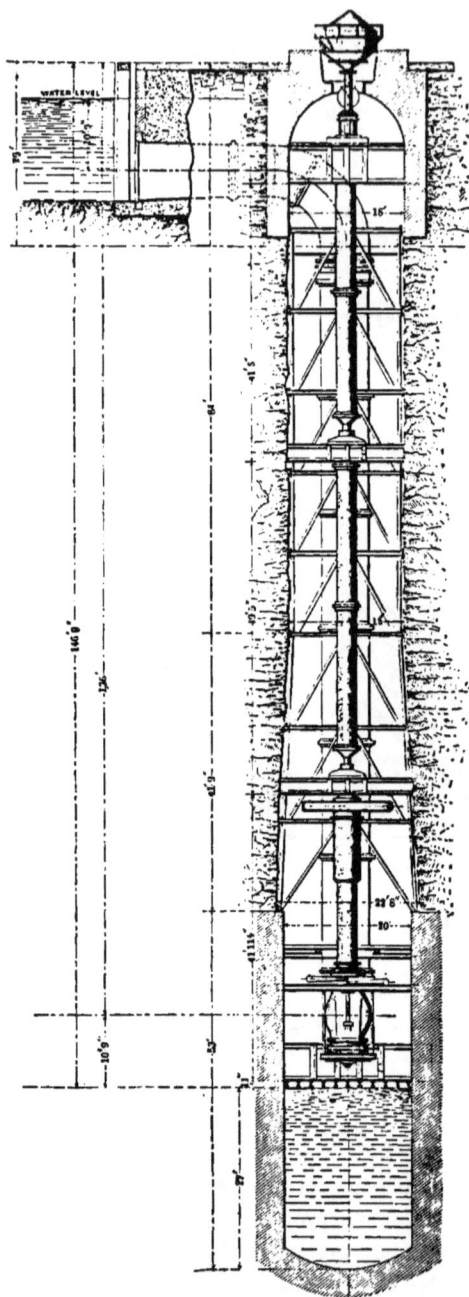

FIG. 81.

Philadelphia, under the superintendence of Messrs. Faesch & Piccard, and they are now being erected.

Hasty critics have assumed that, whatever difficulty there may be about the transmission of the power, the hydraulic part of the problem is of a very ordinary character. That is not so, and the way in which the essential conditions have been met, without sacrifice of efficiency and by arrangements of very great simplicity, reflects great credit on the mechanical skill and judgment of Messrs. Faesch & Piccard. The water descends the wheel-pit to each turbine, fig. 81, in a supply pipe 7 ft. 6 ins. in diameter. The supply pipe bends at right angles at the bottom and delivers the water between a pair of twin outward flow turbines. The hydraulic pressure on the top cover of the wheel chamber, amounting to 60 or 70 tons, is used to support the weight of the vertical shaft and the revolving field magnets of the dynamos, while the pressure on the bottom of the wheel chamber is supported by the foundations. No mechanical arrangement of footstep which could have been designed would have supported so great a weight, on a shaft running at 250 revolutions. One considerable difficulty of the mechanical problem is thus met by means involving absolutely no expense, and perfectly permanent in action. The adjustment of the exact amount of unbalanced upward force can be left to be decided at the last moment. Collar bearings on the shaft support the small difference of load and upward thrust which may unavoidably arise.

The turbine wheels are of bronze, 6 ft. 2 ins. in diameter, and by adopting twin turbines the whole arrangement is made compact, and the double condition of supplying the necessary power at the required speed has been met, without sacrifice of the hydraulic requirements on which the efficiency of the turbines depends. Efficiency at ' part gate' is not important at Niagara, and two ring sluices, on the outside of the turbine wheels, have been adopted for regulating the power. The pressures on these being balanced, they are without difficulty put under the control of a relay speed governor, notwithstanding their great size. To give the governor time to act on the sluices, a fly-wheel of considerable power is necessary. This fly-wheel has been obtained by making the field magnets of the dynamos the revolving part, the comparatively light armature being stationary. The speed regulation of turbines driving dynamos is of the greatest importance, especially as alternating dynamos in parallel are to be adopted. Experience has shown that the speed regulation of turbines of large size by governors presents a mechanical problem of considerable difficulty. The speed governor is a relay governor of a type used by Messrs. Faeseh & Piccard in other installations on a smaller scale. They have so much confidence in its action that they have given a guarantee that when a quarter of the load, that is 1,250 h.p., is suddenly thrown off, the speed variation will not exceed 2 per cent.

The relay is a mechanical relay. A shaft geared to the turbine sluices carries two pulleys, ordinarily running loose and driven by an open and crossed belt. Either pulley can be geared instantly to the sluice shaft by a friction brake, which

x

holds one of a train of three bevil wheels. The brakes are put
in action by a sensitive pendulum governor, acting on ratchet
gear. The governor merely puts the ratchet click in gear.
Hand regulation is provided.

The Electrical Installation.[1]—At the time of the meeting of
the Niagara Commission, the continuous current method of
working was the only one which had been practically used in
any large distribution of motive power. The gradual growth,
subsequently, of a conviction that distant transmission was of
the greatest importance at Niagara led to a decision, in 1893, to
adopt an alternating current system. For the Niagara work,
continuous current methods involved too great a limitation of
the electrical pressure for economical distant transmission. The
alternate current method has one special advantage, in the
possibility of varying the electrical pressure by statical transformers. Initially, the adoption of an alternate current method
was open to the objections that it involves greater stress on the
insulation, and that motors for alternate currents were not
altogether satisfactory for general use, and were almost untried.
Progress in the construction of such motors has been made in
the last year or two, and the objection now has less force. Some
doubt also existed at first as to the satisfactory working of
alternate current generators in parallel. It would be extremely
inconvenient if the generators could not be used in groups
according to the demand for current. Later experience shows
that alternate current generators can be constructed to work
synchronously, in parallel, in a perfectly satisfactory way. The
success of the three-phase motor at Frankfort, and of the two-
phase motors of Tesla in America, has led to the selection of a
two-phase system of working. Such a system with two independent circuits is equally available for working two-phase
and one-phase motors, and for the distribution of part of the
current at any point for ordinary alternate current lighting.
There remain certain applications, such as tramway working,
where continuous currents are most suitable. It appears
probable that methods will be perfected for commutating
alternate currents after their transmission and transformation to

[1] See, for much fuller details, the paper by Professor G. Forbes, ' Electrical
Transmission of Power from Niagara Falls,' *Journal Inst. of Electrical
Engineers.* vol. xxii.

low tension. In any case, motor transformers can be used to obtain continuous currents when necessary.

Acting on the advice of their electrical adviser, Professor G. Forbes, a departure is to be made from previous practice in the electrical system by reducing the frequency of the alternations.[1] Hitherto, alternating currents have been used almost exclusively for lighting, and high frequency (100 to 133 periods) has been commonly adopted, partly to avoid flicker in the lamps, partly because it reduces the cost of transformers. Messrs. Ganz have for some time adopted 42 periods per second. Professor Forbes proposed to use $16\frac{2}{3}$ periods per second; but, after negotiation with the constructors of the dynamos, a frequency of 25 periods has been selected. Professor Forbes believes that low frequency will make parallel working of the dynamos more easy, and improve the efficiency and facilitate the construction of motors. It will reduce the loss due to the unequal distribution of the current in the section of the transmitting conductors, a matter of importance with the large conductors required at Niagara.

Professor Forbes is prepared to construct dynamos to work up to 20,000 volts, in which case step-up transformers would be unnecessary for the transmission to Buffalo. But, proceeding more cautiously, it has been decided that the three first 5,000 h.p. dynamos, now being constructed by the Westinghouse Company, shall work at 2,000 volts, and deliver a two-phase current. For the district at Niagara, electricity will be distributed at 2,000 volts; for transmission to Buffalo it will be transformed to 10,000 volts.

A subway large enough to walk through has been constructed for a third of a mile through the land of the Company. Beyond that the transmission to Buffalo will be by bare copper conductors on insulators carried on posts.

The Scheme on the Canadian side.—The idea of utilising power on the Canadian side was first discussed some years ago. Plans were prepared in which Pelton water-wheels driving dynamos were to be placed in rock chambers underneath the river, just above the Horseshoe Fall, and the tail-water discharged by tunnels, directly on the face of the escarpment forming that fall. Various vague statements were made as to transmitting

---

[1] Mr. Tesla is also an advocate of low frequency for motor work. See *Inventions, Researches, and Writings of Nikola Tesla*; New York, 1894; p. 8.

the electricity to Toronto and Montreal at enormous electric
pressures.

Recently, by charter from the Canadian Government, rights
have been acquired to develop 250,000 h.p. on the Canadian
side, by a company acting in association with the company
developing power on the other side. The Cataract Construction
Company undertakes the engineering work, and is under contract to complete a considerable power development within three
years. The power station will be within the Victoria Park
Reservation, and as at present planned will be similar to the
power station on the other side. The conditions for developing
the power on the Canadian side are more favourable than on the
American side. Deep water in the river comes close to the
shore, and the head-race and tail-race tunnels will be much
shorter. Two tunnels are contemplated, each large enough for
125,000 h.p., and the first of these does not need to be more
than 400 feet in length.

[NOTE.—Since this chapter was in type a paper on the cost of power at
Niagara, by Messrs. E. J. Houston and A. E. Kennelly, has been received.
Allowing 10 per cent. interest and depreciation on the hydraulic works and
turbines, they estimate the cost of power at Niagara at 7s. per effective h.p.
per annum, if working constantly at full load ; or 12s. per h.p. per annum if the
average load is 60 per cent. of the maximum. Then, assuming a three-phase
transmission at a maximum pressure of 50,000 volts, and allowing interest on
dynamos, transformers, motors, and line, and also for loss in transmission,
they calculate the cost of the power delivered at various places. In all cases
the average load is taken at 60 per cent. of the maximum. At Buffalo, 15
miles distant, they estimate the power delivered will cost 2l. 13s. per h.p. per
annum. At Albany, 330 miles distant, they estimate the cost at 5l. 10s. per
h.p. per annum.]

INDEX

ABRAHAM'S air meter, 194
Accumulator, 69; electric, 71; hydraulic, 91, 130, 142
Air compressors, 164, 171; water injection, 177; compound, 179; action in, 223
Air mains, losses in transmission, 184; forms of, 186; cost of, 187; joints for, 187; loss of pressure in, 211; in Paris, 201; flow of air in, 212; experiments on Paris, 219; calculation of, 237
Air motors, 188, 202, 232; at Seraing, 196; action of, 232; experiments on, 235
Alternating currents, 268, 306
Alternating current systems, 283, 294
Antwerp, Compagnie hydro-électrique, 160

BACK pressure in steam engines, 41
Battery storage, 71
Bellegarde, 82, 113, 124
Berne compressed air tramways, 204
Biberist electric transmission, 294
Birmingham, coal consumption in small engines, 28; hydraulic system, 146; air mains, 187; compressed air system, 166; compressors, 227
Boilers, evaporative power, 21; waste in irregular working, 52
Boudenoot vacuum system, 164, 199
Bradford electric station, 65
Bramwell, Sir F., 19

CALORIFIC value of fuels, 10; of ashbin refuse, 16; of gas, 259
Canadian scheme, 307
Carels Frères, Antwerp, engines, 102
Carnegie on natural gas, 257
Castermans on Herstal factory, 287
Cederblom, Prof., on steam turbine, 57

Central stations, advantages of, 4
Coal, calorific value of, 10
Coal consumption in engines, 23; in lighting stations, 27; in small engines, 28
Colladon compressor, 156, 164, 178, 226
Compressed air, 163; history of, 163
Compressed air storage, 59, 197
Compression, isothermal, 173; adiabatic, 174
Compressors, 177; compound, 179, 228; at Paris, 203; action of, 223
Condensation in cylinders, 29
Conductors of electricity, 266
Continuous current systems, 280, 287
Corner on compressed air, 70, 197
Cost of power at Geneva, 6; of liquid fuels, 11; of working destructors, 15; of steam power, 58; of engines and boilers, 58; of gas engines, 60; of steam power in central stations, 61; of accumulator batteries, 71; of thermal storage, 74; of water and steam power, 89; of water power at Geneva, 90; of power at Schaffhausen, 121; of pressure water, 140, 143; of power on hydraulic system, 144, 145; of power at Zürich, 149; of power at Geneva, 159; of air mains, 187; of power from compressed air, 194; of steam, 247; of gas, 256; of electric conductors, 270; of electric transmissions, 273; of power at Niagara, 302
Cotterill, Prof., on cylinder condensation, 44
Creusot, engine experiments, 41
Crompton on coal consumption in lighting stations, 27; on load curves, 36
Cylinder walls, action of, 29, 44

DESSAU electric station, 256
Destructors, 13, 78
Distribution of power by compressed air, 163; theory of, 211; practical calculations for, 237
Distribution of power by steam, 245; by gas, 255; by electricity, 263
Dowson gas, 13, 258, 261
Dwelshaüvers Dery, on action of cylinder walls, 31

EFFICIENCY, mechanical, of engines, 39; thermal, of engines, 43; of turbines, 102, 104; of cable transmission, 114; of hydraulic transmission, 136; of air compressors and motors, 176, 223, 231; of compressed air transmission, 185; of electric transmission, 269; of Herstal installation, 288; of vacuum system, 199; of Shone system, 209; of Genoa installation, 294
Ejectors for sewage, 209
Electric central stations, coal consumption in, 28; load curves of, 38
Electric motors, 283, 293, 306
Electric station at Zürich, 149; at Geneva, 158; at Genoa, 290, 295; at Heilbronn, 294; at Telluride, 295
Ellington, pipe joints, 134; efficiency of hydraulic system, 137
Emerson on testing turbines, 87
Emery, Dr., on cost of water power, 81, 89; on steam distribution, 246
Engine efficiency, 23, 29; friction, 40
Evaporative power of boilers, 19
Evershed on Niagara scheme, 298
Ewing, Prof., on steam turbine, 56

FAESCH and PICCARD governor for turbines, 107, 293, 305
Filters, 143
Flow of water in pipes, 133; of air in pipes, 212; of gas in pipes, 259; of electricity, 268
Flywheel storage, 68
Forbes, Prof. G., on destructors, 18; on utilisation of Niagara, 306
Francis, J. B., on water power, 85
Frankfort-Lauffen experiment, 3, 264
Fribourg rope transmission, 122
Fryer's destructor, 14
Fuel, solid, 9; liquid, 11; gaseous, 12, 259; town refuse, 13

GAS for power purposes, 12, 255
Gas engines, 13, 30, 54, 94, 146, 193, 257

Gasholder storage, 68
Gas Light and Coke Company, load curve, 37
Gas motor. See Gas Engine
Geneva power distribution, 6, 82, 93, 149; electric installation, 295
Genoa electric transmission, 290
Girard turbines, 103; pumps, 148, 154, 156
Gokak rope transmission, 124
Governors for turbines, 105, 158, 292, 305
Greathead injector hydrant, 141
Greenside mines electric transmission, 289
Gutermuth, Prof., experiments on reheaters, 193; on Paris air mains, 217; on compressors, 231; on air motors, 235

HALPIN, thermal storage, 72
Hanarte on compressed air, 165; compressor, 178
Heat exchange in cylinder, 30
Heating power of fuel, 9; of town refuse, 16; of gas, 259
Heilbronn transmission, 285, 294
Herschel, C., on Holyoke water power, 86
Herstal small-arms factory, 287
Hillairet, Domène installation, 264
Hirn on superheating, 32
Hirn, C. F., on telodynamic transmission, 108
Holden, use of liquid fuel, 11
Holly, steam distribution, 245
Holyoke, water power at, 85; testing flume at, 87; tests, 103
Horsfall's destructor, 15
Hughes and Lancaster, tramway system, 206
Hull hydraulic system, 139
Hydraulic motors, 95
Hydraulic Power Company, 5; coal consumption of engines, 27; load curves, 36; accumulator, 92; description of works of the, 140; comparison of hydraulic and air systems, 185, 197
Hydraulic storage, 69
Hydraulic transmission, 129; high and low pressure systems, 131; efficiency of, 136; general arrangement of, 138; examples of high pressure systems, 139; examples of low pressure systems, 146

INDICATED effective horse-power, 39
Internal furnace engines, 54

JACKET, steam, 26, 32
Jackets for compressors, 177, 227
Joint for air mains, 187; for hydraulic
    mains, 135

KAPP, G., on rope transmission, 122;
    on electrical transmission, 263
Keep on refuse destructors, 17
Kelvin, Lord, economical section of
    conductors, 271
Kennedy, J., origin of factory system,
    7
Kennedy, Prof., experiments on boilers,
    53
Kraft, J., use of compressed air, 196

LAVAL steam turbine, 56
Lawrence, water power at, 85
Leavitt, water piston compressors, 178
Liverpool hydraulic system, 145
Load curve and load factor, 35
London hydraulic system, 5, 137, 140

MAINS, for water, 133; for air, 184;
    experiments on, 217; for steam,
    249; for electricity, 278
Mair-Rumley, engine trial, 30
Manchester hydraulic system, 146
Mannesman pipes, 135, 162
Marseilles, efficiency of hydraulic
    plant at, 137
Mekarski tramway system, 164, 205
Meters for water, 141; for air, 167,
    194; for steam, 248
Miller, Oskar von, 72
Mines, compressed air in, 163, 197
Moser, H., Schaffhausen works, 116
Motors for air, 188, 232

NEVADA electric transmission, 289
New York steam system, 246
Niagara, 82, 265; utilisation of, 297
Niagara Commission, 265, 299

OBERURSEL rope transmission, 115
Ochta rope transmission, 116
Oyonaz electric installation, 265

PARIS compressed air system, 164,
    185, 200; experiments on air mains
    in, 219; compressors, 227, 229
Parry, J., on hydraulic power, 145
Parsons, steam turbines, 55
Pearsall, H. D., hydraulic compressor,
    183

Petit and Boudenoot vacuum system,
    164, 199
Petroleum engines, 54
Piccard, relay governors, 105, 158,
    293, 305
Pipes for air, 185, 212; joints for,
    187; loss of pressure in, 222, 238
Pipes for pressure-water, 131; friction
    in, 133; strength of, 134; joints
    for, 135; weight and cost, 136;
    steel, 135, 162; joints for, 135
Pipes for steam, 249; variators for,
    250
Pneumatic transmission, 163
Polyphase systems, 284
Popp compressed air system, 164
Portsmouth Dockyard compressed air
    system, 70, 197
Preller on Berne tramways, 204
Pressure engines, hydraulic, 96, 139,
    152, 158
Pröll, combination of gas and air
    motors, 193
Pulleys for rope transmission, 113

RAFFARD coupling, 160
Reheating compressed air, 191, 233
Relay, hydraulic, 157
Reservoir storage, 92, 130, 149, 151,
    156
Reuleaux, Prof., rope transmission,
    113
Riedler, Prof., valves, 162, 183; com-
    pressors, 201, 229; experiments on
    Paris mains, 217; experiments on
    efficiency of compressors, 231; ex-
    periments on efficiency of motors,
    235
Rieter Brothers, brake for turbines,
    107; rope transmission, 110, 115,
    122, 127; turbines, 291
Rigg, A., hydraulic motor, 97
Rope transmission, 108; at Oberursel,
    115; at Ochta, 116; at Schaffhausen,
    116; at Fribourg, 122; at Belle-
    garde, 124; at Gokak, 124; at
    Genoa, 291
Rotary motors at Paris, 190
Rysselbergh, von, hydro-electric sys-
    tem at Antwerp, 161

SANKEY, Capt., steam consumption
    with variable load, 52
Savage on Terni works, 165
Schaffhausen, rope transmission at,
    81, 116; electric transmission, 122
Schmid motors, 96, 158
Schwoerer, E., superheater, 33
Secondary batteries, 71

Shone, I., sewage system, 164, 207
Siemens, A., continuous and alternate current systems, 287; series system, 282
Siemens, Dr. W., on electric transmission, 263
Société Cockerill, use of compressed air, 195
Solignac on compressed air, 168, 185
Steam consumption in engines, 23; curves of, 46; with variable load, 52
Steam distribution in towns, 245
Stockalper, experiments on air mains, 212, 220
Storage of energy, 67, 90, 130
Subways for electric mains, 280, 307
Superheating, 26, 32
Swain, G. F., water power in America, 83

Telluride electric installation, 265, 295
Telodynamic transmission, 108, 126
Terni steel works, 165, 198
Tesla motors, 283, 296
Testing flume at Holyoke, 87
Thermal storage, 72; cost of, 74; use with destructors, 78; use at gas-generating stations, 78
Thomson, Prof. J., turbine, 103
Thurston, Prof., 87
Thwaite on gas motive-power stations, 4
Town refuse as fuel, 13
Tramways at Berne, 204; Hughes and Lancaster system, 206; electric, 266
Trewby on gas consumption, 255

Turbine, steam, 55
Turbines, 99; at Schaffhausen, 118; at Gokak, 126; at Zürich, 148; at Geneva, 154; at Nevada, 289; at Genoa, 291; at Telluride, 295; at Niagara, 303
Turrettini, Th., Geneva system, 6, 152; hydraulic relay, 157

United States, water power of, 83

Vacuum method of transmitting power, 164, 199
Variators for steam mains, 250

Water power, 80; in United States, 83; relative cost of water and steam power, 89; at Zürich, 147; at Geneva, 149; at Antwerp, 160; at Schaffhausen, 116; at Fribourg, 122; at Bellegarde, 124; at Gokak, 124; in Nevada, 289; at Genoa, 290; in Colorado, 295; at Niagara, 297
Watson on refuse destructors, 14, 17
Weissenbach, on water power, 81
Willans engine, 48
Willans' law of steam consumption in engines, 50
Wind power, 8
Witz, on gas engines, 12; on action of cylinder walls, 30

Zeuner, Prof., tests of turbines, 103
Ziegler rope transmission, 111, 114
Zürich, 82; power transmission at, 146

PRINTED BY
SPOTTISWOODE AND CO., NEW-STREET SQUARE
LONDON

www.ingramcontent.com/pod-product-compliance
Lightning Source LLC
Chambersburg PA
CBHW021503210326
41599CB00012B/1119